BUTTERFLIES
of Louisiana

BUTTERFLIES
of Louisiana

A GUIDE TO
IDENTIFICATION
AND LOCATION

CRAIG MARKS

LOUISIANA STATE UNIVERSITY PRESS
BATON ROUGE

Published by Louisiana State University Press
Copyright © 2018 by Louisiana State University Press
All rights reserved
Manufactured in the United States of America
LSU Press Paperback Original
First printing

Designer: Barbara Neely Bourgoyne
Typeface: Whitman
Printer and binder: Sheridan Books

The base map for all interior maps was created by Mary Lee Eggart.

Library of Congress Cataloging-in-Publication Data
Names: Marks, Craig, 1955– author.
Title: Butterflies of Louisiana : a guide to identification and location / Craig Marks.
Description: Baton Rouge : Louisiana State University Press, [2018] | "By my calculation,
    there are 154 species of butterflies and skippers reported in Louisiana, with the latest
    addition, the Bordered Patch, occurring in October of 2012. Of the 154 total, there are
    8 swallowtails, 5 whites, 13 sulphurs, 1 harvester, 14 hairstreaks, 9 blues, 1 metalmark,
    2 danaids, 1 snout, 3 longwings, 3 fritillaries, 30 brushfoots, 41 grass skippers, 21
    spread-wing skippers, and 2 giant skippers."—Chapter 1. | Includes bibliographical
    references and index.
Identifiers: LCCN 2017037198 | ISBN 978-0-8071-6870-7 (pbk. : alk. paper)
Subjects: LCSH: Butterflies—Louisiana—Identification. | Lepidoptera—Louisiana.
Classification: LCC QL551.L68 M37 2018 | DDC 595.78/909763—dc23
LC record available at https://lccn.loc.gov/2017037198

CONTENTS

PREFACE

Since I was a child, a strong love of nature has dominated my life. I grew up in the unincorporated neighborhood of Whitehaven, south of the city limits of Memphis, Tennessee, and just north of the Mississippi state line. My mom and dad had moved to Memphis in the late 1950s from central Oklahoma for my dad's work. My dad was a farm boy who loved being outdoors fishing, hunting, and hiking. In fact, while attending Oklahoma A&M (now Oklahoma State), he studied to be a forest ranger, but marriage and then two sons required a change in those plans.

So, while the realities of life sent him to Memphis, working as a sales representative for 3M, he never lost his love for the outdoors. As a result, it wasn't long before first he, and then my brother, Gary, and I with him, were exploring locations in the Memphis area where we could experience nature and share that love of the outdoors. I grew up listening to my father's stories of hunting and fishing, through which he taught my brother and me to respect our environment and its nonhuman inhabitants. In my family, hunting outings were a given. However, unlike my father, brother, and ultimately my son (all expert hunters), I became almost exclusively a hunter with a net, looking for butterflies.

By the time I entered high school, chasing butterflies was replaced by sports and associated social activities. I still noticed them, but that was it. Like my collection of baseball cards and old Spiderman comic books, the few boxes of mounted butterflies that survived went into the attic—out of sight and, for the most part, out of mind.

Fast-forward more years than I wish to admit, and transfer my residence to Louisiana. Add to the picture a beautiful wife and three extremely active children (all soccer players). After running competitively during the bulk of that period, I decided to train for triathlons, so that swimming and biking would lessen the pounding on my knees. One day I was competing in the Memphis in May Triathlon, running the final leg, a 10K. At about three miles into the run, we went through a stretch of woods, and suddenly I realized a Zebra Swallowtail was flying along beside me, the first one I had ever seen. Despite the urge to veer off into the woods in pursuit of it, I resisted and finished the race, but old memories were stirring. Within a couple of days of returning home to Louisiana, I saw another one. I took it as a sign. That summer I started chasing butterflies again.

This book represents that love of nature and, in particular, of butterflies that my father instilled in me. It is dedicated to my family: my father, Ferrell Marks; my wife, MaryAnn; my son, Brett; and my two daughters, Elyse and Mattie, all of whom have not just tolerated my passion, but have actually joined me at times (to an understandably less fanatical extent), swung a net, and even provided me with some cool butterflies.

I lost my mom in 2013, after a fifteen-year battle with multiple myeloma. While she didn't accompany us on our hunting trips as boys, later she would go when Dad and I took the grandkids for walks in the woods. I have distinct memories of a trip in late summer one year. Dad had driven us to a high ridge overlooking a vast expanse of cultivated bottomland with the Mississippi River beyond. As we stood enjoying the view, we were surrounded by numerous fresh male Tawny Emperors with their dark maroon and iridescent purple coloring. They swirled around our heads, landed on us, and then dashed away again, a whirlwind of constant motion. I can still see my mom standing in the dappled sunlight, her mouth open in amazement, laughing at her grandkids as they tried to induce the Emperors to land on them. This book is also dedicated to her and the profound influence she had on my life.

BUTTERFLIES
of Louisiana

INTRODUCTION

Butterflies are insects. To be even less scientific, they are bugs, like flies, mosquitoes, fleas, beetles, hornets, wasps, and the other "creepy-crawlies" that inhabit our world. But, for most people, a butterfly is not just another bug. First, through their ability to fly, they both literally and figuratively rise above the creepies and the crawlies. Humans have long been fascinated with flight, and whether it be the long-range flights of the Monarch migrating back to the mountains of Mexico or the short flight of an Eastern Tiger Swallowtail between flowers in a garden, their ability to fly attracts our attention.

However, flight alone does not set them apart because, after all, flies and mosquitoes also have wings. It is their benign existence that truly distinguishes butterflies. In a world full of flying insects that bite, sting, and/or pester, they do none of the above. Their purpose appears to be a pleasant interlude, a momentary reminder that nature can be gentle, nonthreatening, and beautiful.

And there lies the third reason that butterflies have, in the minds of most people, risen above the level of simple "bug" and attained something of an elevated, and certainly cherished, status. Simply stated, they present to us on a regular basis one of the most common examples of nature's beauty. They have been described as flying flowers, combining the fascination of flight with the appeal of color, pattern, and design.

I don't mean to imply that only butterflies among the insect world reflect nature's beauty. But no other group of insects has seemed to capture the same level of appreciation. There are some stunningly beautiful moths. The Luna Moth, for example, can match any butterfly in allure. Moths are, however, primarily creatures of the night, less seen and less followed. There are many very brightly colored beetles capable of flight, albeit not as graceful as butterflies. But, for reasons that I personally cannot explain (or justify), for the most part beetles just don't get much respect from the general populace.

Dragonflies (and their cousins, damselflies) have, of late, garnered growing attention among both birders and butterfliers. Certainly, the Roseate Skimmer, with its hues of blue, pink, and purple, would turn anyone's head for another look. I can attest to both the grace and the beauty of the Ebony Jewelwing, Louisiana's largest damselfly, as it bounces in and out of the shadows of the deep woods where it lives alongside Southern and Creole Pearly-eye butterflies. But dragonflies and damselflies are of a lesser evolved order (Odonata). They appear much more primitive as well as fierce. They are, by their nature, predators and, as such, are apt to dart out of the sky like a military attack airship, capture, and then consume a butterfly that a moment before we were admiring.

And so, we watch butterflies. We chase after them with nets as children (and some continue to do so as grown men). We collect them, photograph them, list them, name them, and, with possibly our most lasting tribute, write about them.

THIS BOOK'S PURPOSE

This book was written to help the reader identify and find butterflies and skippers in Louisiana. With it in hand, someone sitting in a garden can determine what species he or she is observing. Conversely, someone can use it to search for a specific species in likely locations. The book can also serve as a reference on field trips anywhere in the state.

The last annotated list of Louisiana's butterflies and skippers was published in 1965. That list was, in fact, a supplement to an earlier list published in 1954. In 1972, the two referenced lists (and a third, published in 1963) were again supplemented as part of an oral presentation with an unpublished manuscript produced as part of that effort. In the forty-plus years since, much has been learned about Louisiana's butterflies, not only greatly expanding the ranges of those species previously identified, but also identifying additional species. It seemed time to generate an updated list, not one that simply supplemented previous lists but one that combined those prior lists with contemporary data. This book was written to accomplish that purpose.

Louisiana offers numerous opportunities to novices as well as experienced butterfly watchers. Its flight season can stretch from late February to early November, and its multiple habitats include coastal prairies, swamps and bayous, bottomland hardwood forests, and upland piney woods. Common species like the brightly colored and showy Eastern Tiger Swallowtail, Cloudless Sulphur,

Little Yellow, Common Buckeye, and Gulf Fritillary can probably be seen on the wing in every parish. With a little more effort, equally beautiful species like the Silver-spotted Skipper, Pipevine Swallowtail, Monarch, Red Admiral, and Red-spotted Purple can be watched in their natural environment at Louisiana's numerous state and federally protected enclaves. And for those who want a challenge, rarities like the Frosted Elfin, King's Hairstreak, Little Metalmark, and Bay Skipper can be found through careful planning.

Butterflies of Louisiana is a guide to butterflies and skippers that have been, and might again be, found within the Pelican State. I hope it serves you well in identifying and finding the 154 species that grace our region.

BUTTERFLY WATCHING

Butterfly watching has increased in popularity over the last quarter-century, with many more people taking to the field with cameras and binoculars rather than nets. I do not typically carry either of the former items, and so I cannot fairly comment on their proper use or make recommendations on what products to use.

Butterfly watching can be passive (see the comments on butterfly gardening below) or active. That said, there are many LA species that will probably never show up in a garden. For those readers that choose to be active, to go out and seek butterflies rather than wait for them to come to you, keep in mind the concept that butterflies seek mates, propagate, feed, and otherwise exist in each species' own particular "niche" (Pyle 1974). Each species will best be found in those locations that best support that species' biological and environmental requirements. Some species are niche exclusive. For example, the four cane feeders present in LA (Southern Pearly-eye, Creole Pearly-eye, Yehl Skipper, and Lace-winged Roadside-Skipper) are closely tied to their larval food plant, cane. In turn, cane requires very specific environmental conditions (moist, deciduous forest). As a result, those four species are extremely restricted in where they will be found. Louisiana's three elfins also exhibit a close association with their limited larval food plant options along with an association with the habitat those plants require in order to exist in sufficient quantities to support the butterfly's continued reproductive process.

In contrast, some species have very general niches. Painted Ladies and Gray Hairstreaks are good examples. Both are such generalists in their food plant requirements that they are able to successfully reproduce in multiple, environmentally different habitats. Other species, while more selective in their larval food plant requirements, are extremely mobile, moving back and forth between those habitats where their larval food plant is present and other diverse habitats, including gardens in metropolitan areas. Monarchs and Gulf Fritillaries are good examples of these mobile species.

Based on many years of experience, I can make some suggestions to improve the experience of studying butterflies. Butterflies are highly reactive to movement, so go slow and limit extraneous motion. Butterflies can see color. A friend of mine has a shirt with a lot of red on it, and on several occasions Cloudless Sulphurs have closely approached her to investigate her shirt. I wear drab, darker clothing including camouflaged shirts and pants to better blend in with the background. Except for a few satyrs, butterflies cannot hear, so talking should not startle them; however, they are sensitive to vibration, so, again, move slowly and lightly.

As in real estate, one critical key to successful butterflying is location. I have found that butterflies frequent tree lines, roads through wooded regions, roadside ditches, pipelines, and power-line cuts. Look for areas with nectar sources such as patches of thistle, ironweed, Brazilian vervain, boneset, blooming blackberry vines, verbena, asters, and frog fruit. Remember, butterflies are highly mobile, so periodically recheck the same areas during the day. Also keep in mind that many butterflies are associated with and/or attracted to wet areas such as the edges of lakes and ponds, banks of streams and creeks, and even puddles in dirt or gravel roads.

Another important factor is timing. Butterflies require the sun's heat to facilitate flight. If the day is very sunny, butterflies will be on the wing even when temperatures are in the upper 60s, but my rule of thumb is to start looking after the outside temperature has reached 70 degrees. In the Deep South, butterflies are active even if it is overcast, as long as the temperatures are warm enough. Generally, during our Louisiana summers, butterflies will start moving about 9:00 a.m. and continue to fly until as late as 6:00 p.m. However, during the extreme heat of the day, many will limit their movement and seek shade.

Butterfly watching offers an opportunity to monitor our environment. Because butterflies are sensitive to environmental and habitat changes, they provide an excellent measure of the health of a particular location and/or region. By monitoring butterfly populations over the course of a season and then over the course of years, the health of a particular location in the context of outside influences (like human activities) can be evaluated. Downward butterfly population trends could be evidence of developments that in the long run will affect other flora and fauna. Organized butterfly-watching programs can facilitate the generation of curative efforts toward habitat rehabilitation.

BUTTERFLY GARDENING

Another activity that has increased tremendously in popularity in recent years is butterfly gardening, which is done with the goal of attracting to one's yard or garden various species of butterflies rather than going out into the field to search for them. Much of the knowledge I have developed about butterfly gardening came from other sources, and I would strongly recommend three of them. Ross (1994) has written a book specifically related to butterfly gardening in Louisiana. Ajilvsgi (1990) has generated a comprehensive book on this subject for the South in general, as well as one directed to butterfly gardening in Texas. The Minnos (1999) produced a book that addresses butterfly gardening in Florida. These books, and others, are also good sources of information about the preferred food plants of many of the butterflies and skippers to be expected here in LA.

I have built and maintained a couple of butterfly gardens over the last twenty years, all in the Acadiana region of southwestern Louisiana, and can make some recommendations based on what I have found works and doesn't seem to work, at least in that region. Start with a generous amount of lantana, both the low-growing ground cover and the taller bushes. (See appendix B for the scientific names of plants referenced in the text.) If planted in sufficient sun (particularly the afternoon sun), the bush variety will grow as tall as five to six feet and, with Louisiana's long growing season, can be shaped into a hedge line. Lantana comes in various colors, but I found the yellow ground cover and the yellow/pink (called "ham and eggs") and yellow/red bush varieties attract the most attention.

Butterfly bush, particularly the purple variety, is attractive to larger butterflies like swallowtails. I find that I have to regularly replace these bushes as the root systems tend to rot with all of the rain that falls in this region. Purple ironweed does well in my garden and is a native wildflower that attracts a large variety of butterflies and skippers. Another native (and blue/purple) wildflower I use is pickerelweed, but that plant likes its feet to be at least moist. I also have had success planting porterweed (the purple variety works best), but our occasional freezes will kill it unless I bring it into my greenhouse.

I have had little luck with common butterfly weed unless I plant it in a container with a more sand-based soil. Mexican or tropical milkweed does well here and attracts not only monarchs, but also swallowtails and sulphurs. I have also successfully transplanted aquatic milkweed. Both are used by monarchs and queens as a larval food plant.

The larger sulphurs are attracted to red flowers, and I use pentas, ixora, and canna to bring them into the garden. I plant pentas and ixora in pots so they can be brought into the greenhouse during the winter. Even with that precaution, my pentas don't always survive the freezes. I have planted a couple of cassia trees, which produce a yellow flower in the fall and on which Large Orange and Orange Barred Sulphurs will lay their eggs. Although our occasional freezes will kill it back, it has always returned to some degree the following summer.

Other plants that are easy to raise and which are used by certain butterflies as a larval food plant are common passion vine (Gulf Fritillaries) and Dutchman's pipe (Pipevine Swallowtails). I promise that, if you plant dill and/or fennel, you will eventually have Black Swallowtail caterpillars on those plants. Also, if you have room for a small citrus tree, it together with lots of lantana will give you an excellent chance of having your garden visited by Giant Swallowtails.

LIFE CYCLE

All insects go through metamorphosis to reach adulthood. Butterflies and skippers go through complete metamorphosis, which entails four stages: egg, larva (caterpillar), pupa (chrysalis), and adult. The Scott and Cech/Tudor books (see References) have excellent discussions on metamorphosis, as does Matthew Douglas's book, *The Lives of Butterflies* (1986).

Variegated Fritillary egg

Red-spotted Purple egg

Butterflies and skippers lay eggs in a variety of ways. If you watch butterflies long enough, eventually you will see a female flit from plant to plant, actually "tasting" each plant on which she lands with her feet, looking for the correct larval food plant on which to deposit her egg. Most lay them singly, on or near the larval food plant used by that species. In many cases, the eggs are laid on the underside of a leaf or blade of grass or at the base of a leaf or blade. Certain hairstreaks lay their eggs in new flower buds. Fritillaries, which use violets, typically lay their eggs in the area of a violet plant, on the ground or in the surrounding grass. Some brushfoots lay them in rows, others in clusters. The eggs come in various shapes, sizes, and colors. Many have intricate designs and patterns. As the egg matures, it will change color. Upon hatching, the first thing some caterpillars do is to eat the eggshell. The length of time required for an egg to hatch varies depending on species and season. For example, certain species overwinter in the egg stage and do not hatch until the following spring. Other species lay their eggs in early spring, and they mature rapidly, hatching in a matter of days and thus allowing the caterpillars to feed on spring blooms or new growth.

Pipevine Swallowtail eggs

Falcate Orangetip egg

Giant Swallowtail caterpillar

Frosted Elfin caterpillar

As with eggs, caterpillars come in numerous shapes and colors. The primary function of caterpillars is to eat. As they feed, they will go through stages of growth called instars. At the end of each instar, the caterpillar will shed its skin (similar to when a snake sheds its skin), exposing a new, looser skin underneath to facilitate further growth. Since caterpillars are not particularly mobile, their methods of protection from predators include deception, camouflage, chemicals, and living in communal silk nests. Some are camouflaged to look like bird droppings. Others are cryptically colored to match the plant on which they feed. Some swallowtail caterpillars appear to be imitating a centipede, while hairstreak caterpillars are sluglike. Many brushfoot caterpillars are covered with spines or spikes. In some instances, tiny droplets of what are thought to be protective chemicals rest atop bristles or hairs on the caterpillar's body. Certain swallowtail caterpillars possess a Y-shaped fleshy gland behind their head (called an osmeterium), which is everted to emit a protective chemical when the caterpillar is disturbed. Those that live in communal silk nests respond to disturbances by jerking and/or thrashing in unison. I must admit I've never gotten very good at finding caterpillars. On a couple of occasions, an irruption of such proportions occurred that even I noticed the large numbers of caterpillars present (see Question Marks and Buckeyes species accounts).

Gulf Fritillary caterpillar

Pipevine caterpillar

Most butterflies go through five instars during the larval stage, which varies in length according to species. Some overwinter in the larval stage, surviving the colder weather in a sort of suspended-animation state. During the heat of the summer, some of the multi-brooded species complete this stage in two to three weeks. After the fifth instar, the caterpillar typically wanders away from the larval food plant and goes through a transformation process during which it turns into a chrysalis. It is during the pupal stage that the internal organs of the caterpillar break down to be replaced by the adult's internal system. The head, eyes, mouth parts, antennae, working legs, and wings develop, and the transition from larva to adult is completed.

Most chrysalises are cryptically green or brown to match the habitat, mimicking a dead leaf, a thorn, a piece of bark, and so forth. Some are attached to an object such as a plant, a tree, a fence, or even a vehicle. Others rest in the leaf

Monarch chrysalis

Pipevine Swallowtail chrysalis

litter and/or dead grass just above the ground. A few actually burrow into the topsoil. Just before the adult emerges, the chrysalis will change colors, often becoming clear so that the developing adult can be seen. Some species overwinter in the pupal stage, emerging in early spring so that the adults are ready to lay eggs as their food plant begins to present new growth. The more multi-brooded species complete this stage in a couple of weeks in summer. Upon emerging, the new adult (referred to as "freshly eclosed") will have an enlarged abdomen and small shriveled wings. It will then pump fluid from its abdomen into the veins of its wings until the wings expand to full size. During this process and while the wings are drying, the new adult is extremely vulnerable to predators.

Common Buckeye chrysalis

Sleepy Orange chrysalis

Sleepy Orange drying after eclosure

Predation occurs in all four stages of metamorphosis. Historically, the most common predator was thought to be birds, and while birds no doubt do eat butterflies (in all four stages), my experience suggests birds do not eat nearly as many butterflies as other predators do. The list includes mammals (such as mice and rats), reptiles (including many species of lizards), and amphibians (such as toads and frogs). Of this list, anole lizards may be the most common in LA. As for invertebrates, the list is expansive. I have pulled several specimens out of banana-spider webs and pulled a specimen away from a crab spider. I have seen dragonflies catch and eat a butterfly I kicked up from the grass as I walked past. Praying mantises, ambush bugs, and robber flies all prey on adult butterflies. Certain species of beetles, wasps, and ants use large numbers of caterpillars to feed their young. Ichneumon wasps and Tachinid flies parasitize eggs, larvae, and pupae. Egg, larval, and pupal mortality is also affected by disease related to mold, mildew, bacteria, fungi, and viruses. Faced with all of this adversity, less than 5 percent of eggs reach adulthood.

Butterfly populations are also impacted by weather. Severe droughts can cause adult numbers to drop significantly. Flooding, like the historic Louisiana flood of Aug 2016, has a profound impact on smaller butterflies and grass skippers that live and reproduce at ground level. Hard winters with extended periods of below-freezing temperatures tend to kill all of the southern migrants that have moved into the state during the previous summer and fall. Late freezes or sleet after an early-spring warm spell kills both adults and caterpillars. I have noticed that even severe thunderstorms and hurricanes can have a detrimental impact on adult numbers for several weeks thereafter.

The adult butterfly's primary function is to reproduce, thereby restarting the metamorphosis process. Males seek mates in various ways, including patrolling, perching, and hilltopping. Patrolling involves constant movement through areas a male has determined to be a likely location to find females (that is, patches of the larval food plant or nectar-rich flowers). Some swallowtails (Zebras, Blacks, and Eastern Tigers, for example) fly a repeated route, such as up and down the pipeline cuts at Thistlethwaite Wildlife Management Area. Whites, sulphurs, and some brushfoots and satyrs also patrol, but on a more random basis.

Perching entails identifying a location where females will be present and then remaining at a set place such as an isolated and/or extended leaf, tree branch, or stalk of grass while waiting for a female to pass. Males dart out to investigate anything that passes close to their perch, including potential predators like dragonflies (and even humans) before returning back to the same

perch. Grass skippers, hairstreaks, and some brushfoots (Buckeyes, Viceroys, Red Admirals, and Red-spotted Purples, for example) perch. Hilltopping is more of a phenomenon with western butterflies, which live where there are actual hills, unlike in the flat regions of Louisiana. It involves males perching and patrolling at or along high points waiting for females to move up to those heights to mate before laying their eggs at lower elevations.

The actual mating process varies among families as well as species, but it essentially involves the male identifying and approaching a female and then going through assorted movements (sometimes characterized as a courtship dance) to determine if the female is receptive. In some species, pheromones are used to facilitate the process. If the female is unreceptive, typically she will fly away. In some species (such as some sulphurs, satyrs, and the Common Buckeye), she will fly straight up, high into the air, until the male gives up and flies away. If she is approached while feeding on a flower, an unreceptive female will lift her abdomen straight up into the air, thereby making copulation impossible.

In copulation the male and female become attached at the end of their abdomens, facilitated by claspers at the end of the male's abdomen. They can remain so attached for several hours while the male's sperm package is transferred to the female. While this occurs, they typically remain perched motionlessly in the grass or trees but are capable of flight if sufficiently disturbed. In some species, it is the male that will fly while the female dangles below. In others, the opposite occurs. Males are capable of mating multiple times. In some species, the female will mate only once, storing and using the male's sperm package as she produces eggs. In other species, females may mate more than once.

Adult butterflies live for varying lengths of time, again depending on the species. Typically, males have shorter lifespans than females. Most adult butterflies live no more than a couple of weeks. Some species will live several months, particularly those that overwinter as adults (Mourning Cloaks, Question Marks, and Eastern Commas, for example). Monarchs are generally long-lived, and the fall brood that migrates south to Mexico will also survive the winter. Longwings are also long-lived; the Zebra Longwing is reported to live up to eight to nine months.

The greatest source of danger to butterfly populations across the world, including Louisiana, is human interference. Huge numbers of adult butterflies are killed annually by motor vehicles. Even more damaging is the inability to reproduce associated with habitat destruction/loss. Over the last fifteen years, I have seen several butterfly-rich locations, some with unique species present, lost to housing/apartment developments. Even in protected areas such as national forests, nature preserves, and wildlife management areas, I have seen extensive habitat interference in the form of poorly timed or planned burning, bush-hogging, chemical spraying, and mowing. In these locations, the emphasis is on maintaining the area for hunting, with little thought given to the potentially destructive impact those techniques have on butterfly populations. Spraying to control mosquitoes harms populations of smaller, less mobile butterflies and skippers. Pesticides settle on low-growing flowers and shrubs, where the smaller species perch, feed, and overnight. The chemicals don't kill everything, but they do reduce numbers. There is a noticeable ebb and flow associated with areas that undergo scheduled spraying.

CHECKLIST OF LOUISIANA SKIPPERS AND BUTTERFLIES

By my calculation, there are 154 species of butterflies and skippers reported in Louisiana, with the latest addition, the Bordered Patch, occurring in October of 2012. Of the 154 total, there are 8 swallowtails, 5 whites, 13 sulphurs, 1 harvester, 14 hairstreaks, 9 blues, 1 metalmark, 2 danaids, 1 snout, 3 longwings, 3 fritillaries, 30 brushfoots, 41 grass skippers, 21 spread-wing skippers, and 2 giant skippers. See the Note on Sources to learn more about the genesis of this list.

SKIPPERS (SUPERFAMILY HESPERIOIDEA; FAMILY HESPERIIDAE)

Spread-wing Skippers (Subfamily Pyrginae)
    Silver-spotted Skipper (*Epargyreus clarus*)
    White-striped Longtail (*Chioides catillus*)
    Long-tailed Skipper (*Urbanus proteus*)
    Dorantes Longtail (*Urbanus dorantes*)
    Hoary Edge (*Achalarus lyciades*)
    Northern Cloudywing (*Thorybes pylades*)
    Southern Cloudywing (*Thorybes bathyllus*)
    Confused Cloudywing (*Thorybes confusis*)
    Hayhurst's Scallopwing (*Staphylus hayhurstii*)
    Sleepy Duskywing (*Erynnis brizo*)
    Juvenal's Duskywing (*Erynnis juvenalis*)
    Horace's Duskywing (*Erynnis horatius*)
    Mottled Duskywing (*Erynnis martialis*)
    Zarucco Duskywing (*Erynnis zarucco*)
    Funereal Duskywing (*Erynnis funeralis*)

Wild Indigo Duskywing (*Erynnis baptisiae*)
Common Checkered-Skipper (*Pyrgus communis*)
Tropical Checkered-Skipper (*Pyrgus oileus*)
Common Sootywing (*Pholisora catullus*)

Grass Skippers (Subfamily Hesperiinae)
Swarthy Skipper (*Nastra lherminier*)
Neamathla Skipper (*Nastra neamathla*)
Clouded Skipper (*Lerema accius*)
Least Skipper (*Ancyloxypha numitor*)
Southern Skipperling (*Copaeodes minima*)
Fiery Skipper (*Hylephila phyleus*)
Cobweb Skipper (*Hesperia metea*)
Meske's Skipper (*Hesperia meskei*)
Sachem (*Atalopedes campestris*)
Tawny-edged Skipper (*Polites themistocles*)
Crossline Skipper (*Polites origenes*)
Whirlabout (*Polites vibex*)
Southern Broken-Dash (*Wallengrenia otho*)
Northern Broken-Dash (*Wallengrenia egeremet*)
Little Glassywing (*Pompeius verna*)
Arogos Skipper (*Atrytone arogos*)
Delaware Skipper (*Anatrytone logan*)
Byssus Skipper (*Problema byssus*)
Zabulon Skipper (*Poanes zabulon*)
Broad-winged Skipper (*Poanes viator*)
Yehl Skipper (*Poanes yehl*)
Aaron's Skipper (*Poanes aaroni*)
Palatka Skipper (*Euphyes pilatka*)
Dion Skipper (*Euphyes dion*)
Bay Skipper (*Euphyes bayensis*)
Dukes' Skipper (*Euphyes dukesi*)
Dun Skipper (*Euphyes vestris*)
Dusted Skipper (*Atrytonopsis hianna*)
Pepper and Salt Roadside-Skipper (*Amblyscirtes hegon*)
Lace-winged Roadside-Skipper (*Amblyscirtes aesculapius*)
Common Roadside-Skipper (*Amblyscirtes vialis*)

Dusky Roadside-Skipper (*Amblyscirtes alternata*)
Celia's Roadside-Skipper (*Amblyscirtes celia*)
Eufala Skipper (*Lerodea eufala*)
Twin-spot Skipper (*Oligoria maculata*)
Brazilian Skipper (*Calpodes ethlius*)
Salt Marsh Skipper (*Panoquina panoquin*)
Obscure Skipper (*Panoquina panoquinoides*)
Ocola Skipper (*Panoquina ocola*)

Giant Skippers (Subfamily Megathyminae)
Yucca Giant-Skipper (*Megathymus yuccae*)
Strecker's Giant-Skipper (*Megathymus streckeri*)

BUTTERFLIES (SUPERFAMILY PAPILIONOIDEA)

Swallowtails (Family Papilionidae)
Pipevine Swallowtail (*Battus philenor*)
Polydamas Swallowtail (*Battus polydamas*)
Zebra Swallowtail (*Eurytides marcellus*)
Black Swallowtail (*Papilio polyxenes*)
Eastern Tiger Swallowtail (*Papilio glaucus*)
Spicebush Swallowtail (*Papilio troilus*)
Palamedes Swallowtail (*Papilio palamedes*)
Giant Swallowtail (*Papilio cresphontes*)

Whites and Sulphurs (Family Pieridae)
Whites (Subfamily Pierinae)
Checkered White (*Pontia protodice*)
Cabbage White (*Pieris rapae*)
Great Southern White (*Ascia monuste*)
Florida White (*Appias drusilla*)
Falcate Orangetip (*Anthocharis midea*)

Sulphurs (Subfamily Coliadinae)
Clouded Sulphur (*Colias philodice*)
Orange Sulphur (*Colias eurytheme*)
Southern Dogface (*Zerene cesonia*)

White Angled-Sulphur (*Anteos clorinde*)
Cloudless Sulphur (*Phoebis sennae*)
Large Orange Sulphur (*Phoebis agarithe*)
Orange-barred Sulphur (*Phoebis philea*)
Lyside Sulphur (*Kricogonia lyside*)
Barred Yellow (*Eurema daira*)
Mexican Yellow (*Eurema mexicana*)
Little Yellow (*Pyrisitia lisa*)
Sleepy Orange (*Abaeis nicippe*)
Dainty Sulphur (*Nathalis iole*)

Gossamer-winged Butterflies (Family Lycaenidae)
   Harvesters (Subfamily Miletinae)
      Harvester (*Feniseca tarquinius*)

   Hairstreaks (Subfamily Theclinae)
      Great Purple Hairstreak (*Atlides halesus*)
      Juniper Hairstreak (*Callophrys gryneus*)
      Frosted Elfin (*Callophrys irus*)
      Henry's Elfin (*Callophrys henrici*)
      Eastern Pine Elfin (*Callophrys niphon*)
      Oak Hairstreak (*Satyrium favonius*)
      Banded Hairstreak (*Satyrium calanus*)
      King's Hairstreak (*Satyrium kingi*)
      Striped Hairstreak (*Satyrium liparops*)
      Red-banded Hairstreak (*Calycopis cecrops*)
      Dusky-blue Groundstreak (*Calycopis isobeon*)
      Gray Hairstreak (*Strymon melinus*)
      Mallow Scrub-Hairstreak (*Strymon istapa*)
      White M Hairstreak (*Parrhasius m-album*)

   Blues (Subfamily Polyommatinae)
      Cassius Blue (*Leptotes cassius*)
      Marine Blue (*Leptotes marina*)
      Western Pygmy Blue (*Brephidium exilis*)
      Eastern Pygmy Blue (*Brephidium pseudofea*)
      Eastern Tailed-Blue (*Cupido comyntas*)

Spring Azure (*Celastrina ladon*)
Summer Azure (*Celastrina neglecta*)
Ceraunus Blue (*Hemiargus ceraunus*)
Reakirt's Blue (*Echinargus isola*)

Metalmarks (Family Riodinidae)
Little Metalmark (*Calephelis virginiensis*)

Brushfooted Butterflies (Family Nymphalidae)
Snouts (Subfamily Libytheinae)
American Snout (*Libytheana carinenta*)

Milkweeds (Subfamily Danaidae)
Monarch (*Danaus plexippus*)
Queen (*Danaus gilippus*)

Longwings (Subfamily Heliconiinae)
Gulf Fritillary (*Agraulis vanillae*)
Zebra Heliconian (*Heliconius charithonia*)

Greater Fritillaries (Subfamily Heliconiinae; subgenus *Speyeria*)
Variegated Fritillary (*Euptoieta claudia*)
Diana (*Speyeria diana*)

Crescents and Checkerspots (Tribe Melitaeini)
Bordered Patch (*Chlosyne lacinia*)
Gorgone Checkerspot (*Chlosyne gorgone*)
Silvery Checkerspot (*Chlosyne nycteis*)
Phaon Crescent (*Phyciodes phaon*)
Pearl Crescent (*Phyciodes tharos*)
Texan Crescent (*Anthanassa texana*)

True Brushfoots (Subfamily Nymphalinae)
Common Buckeye (*Junonia coenia*)
White Peacock (*Anartia jatrophae*)
Question Mark (*Polygonia interrogationis*)
Eastern Comma (*Polygonia comma*)

Mourning Cloak (*Nymphalis antiopa*)
Milbert's Tortoiseshell (*Nymphalis milberti*)
Red Admiral (*Vanessa atalanta*)
Painted Lady (*Vanessa cardui*)
American Lady (*Vanessa virginiensis*)

Admirals and Relatives (Subfamily Limenitidinae)
Red-spotted Purple (*Limenitis arthemis astyanax*)
Viceroy (*Limenitis archippus*)
Common Mestra (*Mestra amymone*)

Leafwings (Subfamily Charaxinae)
Goatweed Leafwing (*Anaea andria*)

Emperors (Subfamily Apaturinae)
Hackberry Emperor (*Asterocampa celtis*)
Tawny Emperor (*Asterocampa clyton*)

Satyrs and Wood-Nymphs (Subfamily Satyrinae)
Northern Pearly-eye (*Enodia anthedon*)
Southern Pearly-eye (*Enodia portlandia*)
Creole Pearly-eye (*Enodia creola*)
Appalachian Brown (*Satyrodes appalachia*)
Gemmed Satyr (*Cyllopsis gemma*)
Carolina Satyr (*Hermeuptychia sosybius*)
Georgia Satyr (*Neonympha areolatus*)
Little Wood Satyr (*Megisto cymela*)
Common Wood-Nymph (*Cercyonis pegala*)

HISTORICALLY REPORTED BUT QUESTIONABLE

Dina Yellow (*Eurema dina*)
Julia Heliconian (*Dryas iulia*)
Great Spangled Fritillary (*Speyeria cybele*)
Persius Duskywing (*Erynnis persius*)
Leonard's Skipper (*Hesperia leonardus*)

Common Streaky-Skipper (*Celotes nessus*)
Dotted Skipper (*Hesperia attalus*)

After a review of the records and researching each species in the context of its established range outside of Louisiana, I would characterize 9 species as extremely rare and/or accidentally introduced, and another 6 as misidentified/mistakenly included for this state. The former list includes the Byssus Skipper, Polydamas Swallowtail, Florida White, White Angled-Sulphur, Lyside Sulphur, Marine Blue, Diana, Bordered Patch, and Milbert's Tortoiseshell; the latter list includes the Julia, Great Spangled Fritillary, Persius Duskywing, Leonard's Skipper, Common Streaky-skipper, and the Dotted Skipper. The Marine Blue and Lyside have both been seen a total of three times, and the Byssus Skipper, Polydamas Swallowtail, and Florida White have each been seen twice. The others have been reported only one time within the state.

Another 10 should be described as strays, some of which occasionally appear to colonize within the state only to later disappear. These species include the White-striped Longtail, Cabbage White, Large Orange Sulphur, Orange-barred Sulphur, Mexican Yellow, Mallow Scrub-hairstreak, Zebra Longwing, White Peacock, Common Mestra, and Appalachian Brown.

The remaining 129 appear to be permanent residents of Louisiana to some degree or another; however, I would note that one of these, the Yucca Giant-Skipper, may now be on the verge of extirpation within the state. I also have questions about the current status of the Gorgone Checkerspot. Although reported in numbers as recently as the 1990s as part of the survey of the Cajun Prairie, only one has been seen since. I have similar concerns about the Cobweb Skipper and Dotted Skipper (if the latter was ever actually present within the state) as it has been even longer since either of these species has been reported within the state.

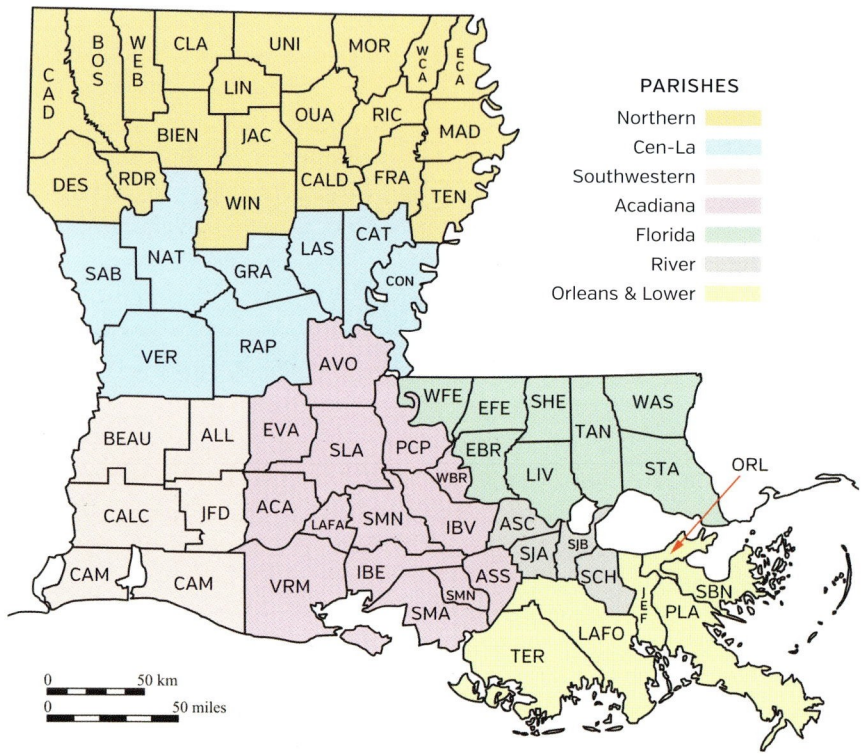

PARISHES

Northern
Cen-La
Southwestern
Acadiana
Florida
River
Orleans & Lower

0   50 km
0   50 miles

Parishes and Regions of Louisiana

WHERE THE BUTTERFLIES ARE IN LOUISIANA

BY PARISH

Louisiana is subdivided into governing units called parishes rather than counties. There are sixty-four parishes in the state. These are listed below, together with the abbreviations used for these parish names in the species accounts. The number in parentheses beside each parish name indicates the total number of species reported to date in that parish. I have also included within each species account a distribution map that reflects the parishes where the species has been sighted.

ACA	Acadia Parish (40)	EBR	East Baton Rouge Parish (88)
ALL	Allen Parish (52)	ECA	East Carroll Parish (20)
ASC	Ascension Parish (19)	EFE	East Feliciana Parish (87)
ASS	Assumption Parish (11)	EVA	Evangeline Parish (58)
AVO	Avoyelles Parish (40)	FRA	Franklin Parish (13)
BEAU	Beauregard Parish (50)	GRA	Grant Parish (76)
BIEN	Bienville Parish (55)	IBE	Iberia Parish (50)
BOS	Bossier Parish (87)	IBV	Iberville Parish (39)
CAD	Caddo Parish (106)	JAC	Jackson Parish (33)
CALC	Calcasieu Parish (41)	JEF	Jefferson Parish (67)
CALD	Caldwell Parish (49)	JFD	Jefferson Davis Parish (36)
CAM	Cameron Parish (73)	LAFA	Lafayette Parish (75)
CAT	Catahoula Parish (63)	LAFO	Lafourche Parish (47)
CLA	Claiborne Parish (69)	LAS	LaSalle Parish (56)
CON	Concordia Parish (37)	LIN	Lincoln Parish (40)
DES	DeSoto Parish (79)	LIV	Livingston Parish (50)

MAD	Madison Parish (30)	SLA	St. Landry Parish (80)
MOR	Morehouse Parish (31)	SMA	St. Mary Parish (50)
NAT	Natchitoches Parish (103)	SMN	St. Martin Parish (61)
ORL	Orleans Parish (80)	STA	St. Tammany Parish (110)
OUA	Ouachita Parish (36)	TAN	Tangipahoa Parish (84)
PLA	Plaquemines Parish (20)	TEN	Tensas Parish (27)
PCP	Pointe Coupee Parish (38)	TER	Terrebonne Parish (65)
RAP	Rapides Parish (95)	UNI	Union Parish (28)
RDR	Red River Parish (43)	VER	Vernon Parish (88)
RIC	Richland Parish (23)	VRM	Vermilion Parish (45)
SAB	Sabine Parish (87)	WAS	Washington Parish (49)
SBN	St. Bernard Parish (19)	WBR	West Baton Rouge Parish (32)
SCH	St. Charles Parish (51)	WCA	West Carroll Parish (22)
SHE	St. Helena Parish (65)	WEB	Webster Parish (56)
SJA	St. James Parish (16)	WFE	West Feliciana Parish (95)
SJB	St. John the Baptist Parish (49)	WIN	Winn Parish (61)

Some parishes are badly underreported, as evidenced by comparing their numbers with those of nearby parishes. ASC, ASS, and FRA all have fewer than 15 species recorded within them. Thirteen others have fewer than 35 recorded species. Only 3 parishes, STA, CAD, and NAT, have over 100 species recorded—110, 106, and 103 respectively. Two more parishes have 90 or more reported, WFE (95) and RAP (95).

No butterfly or skipper has been reported from all 64 parishes to date. The species reported in the most parishes are the Cloudless Sulphur and Little Yellow, both in 62 parishes. The swallowtails, the sulphurs, and the brushfoots seem to be the most reported families.

The top two spread-wing skippers were Silver-spotted Skippers and Horace's Duskywings, both in 49 parishes. The most-reported grass skipper was the Fiery Skipper, found in 54 parishes. Other notable grass skippers were the Clouded Skipper (49), the Least Skipper (45), and the Southern Skipperling (44). The Tiger Swallowtail (57) and Pipevine Swallowtail (57) were the top-recorded swallowtails. The statistics for whites are misleading because, while the Checkered White (34) has shown up in more parishes than the Great Southern White (23), I feel the latter is the most common white in the state. The same is true for the sulphurs. The Cloudless Sulphur and Little Sulphur have

been seen in the same number of parishes (62), but I believe the latter is more numerous across the state.

The top two hairstreaks were the Red-banded Hairstreak (54) and the Gray Hairstreak (52), and the top two blues were the Eastern Tailed-Blue (39) and the Summer Azure (22). For the brushfoots, the two reported in the most parishes were the Pearl Crescent (60) and Common Buckeye (58). Monarchs have been reported in 57 parishes. Other notables included the Gulf Fritillary (57), the American Snout (48), the American Lady (51), the Viceroy (52), and the Question Mark (52). The most reported satyr/wood nymph was the Carolina Satyr, in 55 parishes.

BY REGION

In describing where each species has been found, I refer in this guide not only to specific parishes, but also to regions into which the parishes have been grouped. Those regions are north Louisiana, central Louisiana, Acadiana, southwest Louisiana, the River Parishes, the Florida Parishes, and the New Orleans and Lower Parishes region (see map). These regions are defined and described below. For each, I list places where many of the butterflies referenced in the species accounts have been seen. In addition to postal abbreviations for states, the following abbreviations are used in these descriptions and the species accounts:

BAMONA	Butterflies and Moths of North America website, www.butterfliesandmoths.org
HQ	headquarters
LSU	Louisiana State University
NABA	North American Butterfly Association
NF	National Forest
NO	New Orleans
Rec Area	Recreation Area
SF	State Forest
SGCN	Species of Greatest Conservation Need
SP	State Park
NWR	National Wildlife Refuge
WMA	Wildlife Management Area

North Louisiana

North Louisiana is known locally as "Sportsman's Paradise" (a name also attributed to LA as a whole). In general, uplands are dominated by pines, bottomlands by hardwood forests. The predominant pine is longleaf toward the south and shortleaf with a significant hardwood element to the north. This region is primarily rural. As the nickname would imply, the area presents numerous outdoor activities. The rolling pine hills are excellent for hunting deer, wild turkey, and ducks. Abundant waterways offer opportunities for freshwater fishing. There are also a lot of places that are excellent for butterflies. CAD is one of the most extensively covered parishes in north LA. Located in the extreme northwestern corner of the state, it is bordered by Texas on the west, Arkansas on the north, and the Red River on the east. It consists of both upland and bottomland habitats. The upland areas sustain both pine and hardwood forests. The bottomland areas are primarily in the Red River floodplain and harbor several deciduous tree types like hackberry, cottonwood, willow, sycamore, and ash.

PLACES WHERE BUTTERFLIES HAVE BEEN SPOTTED

Bayou Macon Wildlife Management Area, in ECA, approximately 3.5 miles east of Oak Grove and 7.5 miles northwest of Lake Providence.

Bayou Pierre Floodplain, thirty minutes south of Shreveport, off Hwy 1.

Bodcau Wildlife Management Area, in BOS and WEB northeast of Bossier City, off LA Hwy 157.

C. Bickham Dickson Park, 2283 Bert Kouns Industrial Loop within the city of Shreveport.

Caney Ranger District of Kisatchie NF (including Caney Lakes Rec Area, Middle Fork Unit, and Corney Lake Unit) in CLA, northeast of Shreveport and south of the AR state line near Homer (HQ at 3288 Hwy 79, Homer, LA).

Copenhagen Hills Preserve, six miles southeast of Columbia in CALD.

D'Arbonne National Wildlife Refuge, in OUA and UNI, on White's Ferry Road (Hwy 143). The refuge complex HQ is approximately seven miles north of West Monroe at the intersection of Hwy 143 and Holland's Bluff Road.

Driskill Mountain (also referred to as Mount Driskill), eleven miles south of Arcadia, in BIEN, reached via I-20, using the Arcadia exit on State Route 147.

Eddie D. Jones Park, Hwy 789 in southwest CAD.

Gum Springs Recreation Area, west of Winnfield.

Headquarters Unit, accessed from LA 1 in RDR.

Lake Claiborne State Park, at 225 State Park Road, Homer.

Loggy Bayou Wildlife Management Area, in southern BOS, off LA Hwy 154 just east of Lake Bistineau, approximately twenty miles southeast of Bossier City.

Red River National Wildlife Refuge, along the Red River between Colfax and the AR state line.

Tensas River National Wildlife Refuge, in three parishes of northeast LA (MAD, TEN, and FRA), west of the city of Tallulah.

Walter B. Jacobs Nature Park, three miles west of Blanchard, on Blanchard Furr Road.

Winn Ranger District, near Winnfield with 164,000 acres in WIN, NAT, and GRA.

Central Louisiana

Central Louisiana is also known as the Crossroads region. It is a land of piney hill country, extensive prairies, deciduous forests, and some swampland. It also has clear streams with sandy bottoms. Some areas include hills and mesas that qualify as steep and rocky, although none are more than four hundred feet high. Kisatchie National Forest, the only national forest in Louisiana, is spread across seven parishes, many of them in central LA. Kisatchie NF is divided into five managed units that are called ranger districts and that total over 604,000 acres of public lands. For more information, directions, and maps, go to the Kisatchie NF website.

PLACES WHERE BUTTERFLIES HAVE BEEN SPOTTED

Alexander State Forest and **Indian Creek Lake and Recreation Area**, approximately ten miles south of Alexandria, off I-49 South (Exit 73).

Calcasieu Ranger District, including the **Vernon Unit**, within which there are several specific areas worth visiting, including the **Blue Hole Recreation Area**, on Forest Service Road 405 off LA 10.

Catahoula National Wildlife Management Refuge, twelve miles east of Jena.

Catahoula Ranger District, including **Stuart Lake Recreation Complex** (reached via LA Hwy 8, four miles east of Bentley, on Stuart Lake Road [FR

144]) and the **Catahoula Hummingbird and Butterfly Garden** (across the road from the main office on Hwy 8).

Cloud Crossing Recreation Complex, near Goldonna, off Hwy 1233, on Forest Road 513.

Cooter's Bog Special Interest Area (LA Hwy 463), **Leo's Bog** (Gravel Hill Church Road), and **Dove Field** (east off LA 399, one mile north of community of Fullerton). Also within that district is the **Evangeline Unit**, which includes the **Castor Creek Scenic Area** (near Woodworth off Caster Plunge Road) and **Wild Azalea Seep** (Messina Road off LA 488).

Kisatchie Ranger District, including the **Longleaf Vista Recreation Area**, **Kisatchie Bayou Recreation Complex**, and **Kisatchie Hills Wilderness Area**, all accessible from I-49 at the Derry exit, number 119, on Forest Hwy 59.

Lower Cane River, south of Natchitoches in NAT near Melrose Plantation.

Sicily Island Hills Wildlife Management Area, in northeast CAT six miles east of Harrisonburg on Hwy 8.

Spanish Lake Lowlands, north of Natchitoches in NAT, bisected by I-49.

Acadiana

In 1971, the Louisiana state legislature officially designated a 22-parish area of the state as "Acadiana." For the purposes of this guide, however, Acadiana refers to twelve parishes at the heart of Cajun Country. Despite the frequent association of Cajuns with swamplands, Acadiana consists mainly of low gentle hills in the north section and dry-land prairies, with marshes and bayous in the south closer to the coast, increasing in frequency in and around the Atchafalaya and Mississippi basins. Acadiana has an abundance of rice and sugarcane fields. The Atchafalaya Basin is a scenic semi-wilderness area of hardwood forests, cypress stands, marshes, and bayous. It is one of the last great river swamps left in the nation. The area encompasses 28,500 acres and contains levees, forested wetlands, bayous, and shallow lakes. Acadiana also includes an area described as the Cajun Prairie, a type of tallgrass prairie similar in many ways to that of the US Midwest. So named because it was settled in the early nineteenth century by exiled Acadian settlers, the Cajun Prairie originally encompassed as many as 2.5 million acres but has been significantly reduced as the land was used for agriculture. The remaining remnants of the Cajun Prairie contain a high diversity of native tallgrass prairie flora.

PLACES WHERE BUTTERFLIES HAVE BEEN SPOTTED

Avery Island "Jungle Garden," in IBE at end of Avery Island Road (LA 329).

Bayou Teche National Wildlife Refuge, in SMA, southeast of Franklin, near the town of Garden City, with multiple units accessible off LA 318.

Burn's Point Rec Area in SMA on Burn's Point Road off LA 317.

Cajun Prairie Region, including **Eunice Cajun Prairie** (in Eunice, at the corner of Martin Luther King Drive and Magnolia Street) and **Duralde Prairie** (between Eunice and Mamou).

Chicot State Park and the **LA State Arboretum,** within the park, on LA 3042/ Chicot Park Road, about seven miles from Ville Platte.

Cypremort Point State Park, primarily in SMA, along Vermilion Bay on Beach Lane off LA 319.

False River and **Old River**, oxbow lakes in PCP.

Freshwater Bayou Lock Recreation Area, on Route LA 3147 in VRM.

Indian Bayou Wildlife Management Area, in the heart of the Atchafalaya Basin, north of I-10, south of Hwy 90, and west of the Atchafalaya River.

Palmetto Island State Park, in VRM south of Abbeville at 19501 Pleasant Road.

Paul J. Rainey Wildlife Sanctuary, essentially to the north and the west of the end of LA 3147.

Rip Van Winkle Gardens, in IBE at Jefferson Island.

Sherburne Wildlife Management Area, east of Indian Bayou, accessed from the Whiskey Bay and Ramah exits off I-10 in IBV.

Thistlethwaite Wildlife Management Area, in north-central SLA, northeast of Washington and accessible off I-49 at the Lebeau/LA 10 exit.

Southwest Louisiana

Southwest Louisiana is a five-parish area with Texas on its western border and the Gulf of Mexico on its southern. It contains a wide array of habitat types, including prairie, gallery woods, cypress-tupelo forest, hardwood swamps, longleaf pine forest, freshwater and saltwater marshes, cheniers (oak woods on old ridges of the gulf beach), beach, and open gulf. This portion of LA is geographically and culturally attached to southeast Texas. Of note to birders and butterfliers, the Creole Nature Trial through this region is a network of scenic highways and walking paths that wind through 180 miles of marsh, bayou, and shoreline.

PLACES WHERE BUTTERFLIES HAVE BEEN SPOTTED

Cameron Prairie National Wildlife Refuge, along Hwy 27E, north of Creole.

Lacassine National Wildlife Refuge, in JFD, southeast of Welsh with access off LA 14, south on Hwy 3046, then right on Streeter Road.

Lacassine Pool, at the end of Illinois Plant Road off LA 14, south of Welsh.

Peveto Woods, a bird sanctuary owned by the Baton Rouge Audubon Society off Hwy 82, 8.5 miles west of Holly Beach.

Rockefeller National Wildlife Refuge, just east of Grand Chenier, on Hwy 82.

Sabine National Wildlife Refuge, including the **Blue Goose Trail** and the **West Cove Trail**, both along Hwy 27W.

Sam Houston Jones State Park, north of Lake Charles at 107 Sutherland Road.

River Parishes

The River Parishes are located between New Orleans and Baton Rouge along both banks of the Mississippi River. They are traditionally considered part of Acadiana, although for our purposes they are treated as a separate region. St. Charles and St. John the Baptist parishes also make up an area historically referred to as the German Coast, whereas Ascension and St. James parishes were known as the Acadian Coast. Their habitat is primarily moist bottomland hardwood forests, very similar to the habitat found in the eastern portion of the Florida Parishes and the northern portion of the New Orleans region.

PLACES WHERE BUTTERFLIES HAVE BEEN SPOTTED

Bonnet Carré Spillway, in SCH, between the towns of Montz and Norco about thirty-three miles above Canal Street in New Orleans. The project office is in Norco.

Florida Parishes

The Florida Parishes comprise the southeastern part of LA that was part of Spanish West Florida in the early nineteenth century. Unlike the rest of LA, this area remained under Spanish control and was not part of the Louisiana Purchase. The FL Parishes stretch from the Mississippi state line on their eastern and northern borders to the Mississippi River on their western border and Lake Pontchartrain on part of their southern border. The area above Lake Pontchartrain is known as the Northshore.

The region offers excellent examples of pine-dominated wetland communities, including longleaf pine savanna, longleaf flatwoods, bayhead, slash pine/pond cypress woodland, and riparian forest. It also contains other unique habitats such as a bayhead forest, a small river floodplains forest, and upland sandy streams. On the western side, the two Feliciana parishes present terrain that is typified by rugged hills, bluffs, and ravines. Those two parishes lie at the southern end of the Loess Blufflands escarpment that follows the east bank of the Mississippi River. These bluffs offer a diverse habitat that supports some species of plants and animals not found elsewhere in Louisiana. The area is classified as upland hardwood, with some loblolly pine and Eastern red cedar mixed in on the ridgetops and creek terraces. It is open to a variety of outdoor recreational activities, including hunting and trapping.

PLACES WHERE BUTTERFLIES HAVE BEEN SPOTTED

Abita Creek Flatwoods Preserve, east of Abita Springs, on the Talisheek Hwy (Hwy 435) in STA.

Asphodel Plantation, in EFE at 4626 LA 68, Jackson.

Big Branch Marsh National Wildlife Refuge, 58490 Lake Road, Lacombe (HQ at 61389 LA 434, Lacombe).

Bluebonnet Swamp Nature Center, within the city limits of Baton Rouge at 10503 N. Oak Hills Parkway.

Clark Creek Nature Area, not actually in LA, but just over the state line from WFE at Pond, MS.

Fairview-Riverside State Park, at 119 Fairview Drive, Madisonville, two miles east of town.

Fontainebleau State Park, south of I-12 and southeast of Mandeville on US 190.

Global Wildlife Center, in TAN at 26389 Hwy 40, Folsom.

Greenwood Park, in EBR on Hwy 19 near Baker.

Lake Ramsey Savannah Wildlife Management Area, about seven miles northwest of Covington on Lake Ramsey Road west of LA Hwy 25.

Mary Ann Brown Preserve, in WFE between St. Francisville and Jackson, at 13515 Hwy 965.

Myrtles Plantation and **Butler-Greenwood Plantation**, in St. Francisville, in WFE.

Northlake Nature Center, across the highway from Fontainebleau State Park in STA.

Pearl River Wildlife Management Area, in STA approximately six miles east of Slidell, accessible via old Hwy 11.

Tickfaw State Park, in LIV; take the Albany/Springfield exit from I-12 and follow the signs.

Tunica Hills Wildlife Management Area, composed of two separate tracts lying northwest of St. Francisville in WFE. The South Tract is 17.3 miles west of LA Hwy 61 on LA Hwy 66.

Waddill Wildlife Refuge, in Baton Rouge on North Flannery Road.

Weyanoke, an area of hunting leases along Ouida Irondale Road off LA Hwy 61, about three miles past the community of Weyanoke in WFE.

New Orleans and Lower Parishes

The city of New Orleans and the parishes below it make up what we will call the New Orleans and Lower Parishes region. Fourchon is LA's southernmost port, located at the southern tip of Lafourche Parish, on the Gulf of Mexico. This region contains areas of fresh and brackish marshes, coastal hardwood forests, lagoons, swamps, canals, borrow pits, cheniers, and natural bayous. The climate is humid and subtropical. Despite occasional freezing weather, the long growing season, abundant rainfall, and fertile soils support a great variety of plant and animal life—shrimp in the marsh, crawfish in the swamp, and butterflies and orb-weaving spiders in the forest.

PLACES WHERE BUTTERFLIES HAVE BEEN SPOTTED

Barataria Preserve and **Jean Lafitte National Historical Park and Preserve**, in JEF and PLA on LA 45 in Marrero. The visitor center is at 6588 Barataria Boulevard, also in Marrero, near Crown Point.

Bayou Sauvage National Wildlife Refuge, the nation's largest urban NWR, within the city limits of New Orleans, accessed from Hwy 11 and Hwy 90.

Fourchon Beach, west of LA 1 on LA 3090 (also known as Fourchon Road).

Grand Isle State Park, in JEF at Admiral Craig Drive on Grand Isle.

Mandalay National Wildlife Refuge, in TER with its headquarters off LA 182, about five miles west of Houma.

Pointe-aux-Chenes Wildlife Management Area, in southeastern TER, approximately fifteen miles southeast of Houma. Access is available from the Island Road and on Hwy 665 as well as in the Montegut area behind the Montegut Middle School off Dolphin Street.

BY SEASON

Despite a primary flight season from early to mid-March through the end of October (and into early November in some years), butterflies can be seen in all months of the year in LA, particularly in the southern portion of the state. Based on the records used for this book, 29 and 49 species have been reported in the months of January and February, respectively, probably the coldest months of the year. By March, the total number of species reported increased to 84. April records jumped up to 112 species with May's total only 5 less at 107. June records increased to 114, the month with the most species recorded. As the heat and humidity of July and August reduce activity, the number of species dropped slightly to 108 for both. September numbers match July and August, and October's total was only 2 species fewer at 106. By November, while it is still possible to find many butterflies present in large numbers (such as Buckeyes, Gulf Fritillaries, and Ocola Skippers), the total number of species dropped to 83. The total number of species reported in December was 38.

COUNTING AND REPORTING

The Lepidopterists' Society is an international organization that was founded in Cambridge, Massachusetts, in 1947 to promote the scientific study of lepidopterology and facilitate the sharing of specimens and ideas by professionals and amateurs. Among its publications is an annual "Season Summary" that provides reports from members of species found in Canada, Mexico, and the US during the preceding year, including numbers, locations, dates, and host plants. The Season Summary is also viewable in a searchable format on the society's website.

The North American Butterfly Association (NABA), formed in 1992, is a nonprofit organization open to anyone in North America who is interested in butterflies. Its goals are to increase public enjoyment and conservation of butterflies. As part of this effort, NABA oversees the Butterfly Count Program in the US, Canada, and Mexico. It assumed responsibility for the program in 1993 from the Xerces Society, which began the counts in 1975. There are currently about 450 counts, which basically involve one or more individuals (both members and nonmembers) making a list of all butterflies (including eggs, larvae, and pupae) observed within a fifteen-mile-diameter range of a designated site in a one-day period. Midsummer of the year following the count, NABA publishes a bound report that provides the details of each count (what, how many, date). This annual report is helpful in identifying potential geographical distribution and relative population sizes of the species sighted in a given region or state.

Over the years, LA counts have been held at Alexandria, Allen Acres, Barataria, Bonnet Carré Spillway, Cajun Prairie, Cameron, Catahoula Butterfly Garden, Catahoula NWR, Copenhagen Hills, Fort Polk, Folsom, Grand Isle, Indian Bayou WMA, Kisatchie National Forest (NAT), Lower East Pearl River, Metro New Orleans, Shreveport, St. Tammany/Global Wildlife, Thistlethwaite WMA, and west STA. The Lower East Pearl River count is one of three counts nationwide that has been conducted each year since 1975. Part of that count is held in the Honey Island Swamp WMA in STA.

The precise location for each count is described in the annual report. Some locations are self-evident by name; other locations have a more generic name. For example, the Alexandria count range included portions of Kisatchie NF in RAP. Similarly, the Barataria count covered Barataria Preserve. The Cajun Prairie count encompassed the Eunice Prairie and the Duralde Prairie restoration projects. The center of the Cameron count was Peveto Woods Sanctuary near the community of Johnson Bayou. The Catahoula count included the Catahoula Butterfly Garden and Stuart Lake. The Folsom count was centered on Mizell Farms (a plant nursery), in conjunction with a butterfly festival held there each year. The Fort Polk count involved sections of Kisatchie NF in VER and included Cooter's Bog and Blue Hole. The Metro New Orleans count included the Audubon Zoo as well as numerous private gardens. The west St. Tammany count included Lake Ramsey and Abita Springs.

IDENTIFICATION KEYS

SCIENTIFIC CLASSIFICATION

Butterflies are insects. Insects belong to a large group scientifically identified as the phylum Arthropoda (the classic evolutionary group that has been modified in recent decades), which includes not only insects, but spiders, crabs, crustaceans, millipedes, and so forth, all of which have jointed legs, segmented bodies, and an external skeleton. This phylum is divided into classes, with insects forming the class Insecta. The distinguishing features of insects are three major body sections (head, thorax, and abdomen); six legs; and two antennae. The class Insecta is further divided into orders. Butterflies, skippers, and moths belong to the order Lepidoptera (which means "scaled wings"). In terms of evolution, Lepidoptera appears to be one of the most recent insect orders, with butterflies thought to be the farthest along the evolutionary chain, moths being the lowest, and skippers in between.

DISTINGUISHING MOTHS FROM BUTTERFLIES AND SKIPPERS

The order Lepidoptera has been divided into two sections, Microlepidoptera and Macrolepidoptera. The section Macrolepidoptera has been subdivided into Heterocera (moths) and Rhopalocera (butterflies and skippers). There are significantly more moths in LA than butterflies and skippers; however, the majority of LA's moths are nocturnal and not typically encountered while in the field looking for butterflies. Nonetheless, there are some moths that do fly, or can occasionally be found, during the day.

After years of practice, I am still, on occasion, fooled by a day-flying moth. There are several distinctions typically given between moths and butterflies/

skippers, the principal one being the antennae. Moth antennae normally are either thread- or feather-like, while butterfly antennae have a distinctive knob or club at the end. The antennae on skippers usually have clubs that are tapered at the end, hooking backward like a crochet hook.

butterfly

skipper

moth

moth

It is not always easy to get a good look at antennae of small moths in the field, but their flight pattern is another way to distinguish them from butterflies and skippers. Specifically, after having been flushed, rather than land and perch like a butterfly, small moths will abruptly close their wings and simply drop to the ground, into the grass and/or undergrowth, seemingly disappearing from sight. In addition, many (but not all) moths both small and large will hold their wings over their back like a roof, a trait that butterflies and skippers do not share.

Rosy Maple Moth

USING THIS GUIDE TO IDENTIFY BUTTERFLIES AND SKIPPERS

The suborder Rhopalocera has been further divided into two superfamilies: Hesperioidea (skippers) and Papilionoidea (butterflies). Both skippers and butterflies have been further divided into families. To facilitate proper identification of an encountered butterfly or skipper, a series of keys followed by general descriptions and sample photos of each family are provided below. Use these aids to estimate the family of the species in question. Once a family match is made, go to the species accounts for that particular family in this guide. Use the photos to narrow down the species. Then study the entry and distribution map for that species to either confirm identification or to continue the search.

KEYS

Begin by determining or estimating wing size. When the wings are spread open, consider the horizontal distance between left upper-wing tip and right upper-wing tip.

(1) Large wing size (3 inches or more):
  (a) Tails on hindwings: *see swallowtails (p. 41, pp. 190–211).*
  (b) No tails on hindwings: consider wing color:
    (i) Wing color orange/brown: *see brushfoots/Milkweeds (p. 47, pp. 314–17).* Viceroys and Variegated Fritillaries (both brushfoots) also have orange/brown wing color but usually have a wing size smaller than 3 inches.
    (ii) Wing color black/dark blue: *see Red-spotted Purple (pp. 360–62) and Red Admiral (pp. 354–55).* Both are brushfoots.

(2) Medium wing size (1.5 to 3 inches):
  (a) Small, blunt tail projections on the hindwing and/or ragged, rough outer wing edges: *consider True Brushfoots such as the Question Mark (pp. 346–47), Eastern Comma (pp. 348–49), or Mourning Cloak (pp. 350–52); check also the Goatweed Leafwing (pp. 368–69).*
  (b) No tails: consider wing color:
    (i) Wings are white, yellow, or bright orange: *see whites and sulphurs (pp. 41–42, pp. 212–53).*
    (ii) Wings are duller orange/brownish orange: *consider True Brushfoots such as the Viceroy (pp. 363–65) or Variegated Fritillary (pp. 324–25).*
    (iii) Wings are dull gray or brown: *see brushfoots/Satyrs and Wood-Nymphs (pp. 48–49, pp. 374–97).*
  (c) Another factor to consider is wing shape:
    (i) Long slender forewings, flies with shallow wing beats: *see brushfoots/Longwings (p. 48, pp. 318–23).*
    (ii) Specimen is dark with short, triangular forewings and a thick, cigar-shaped body: *see Giant Skippers (p. 40, pp. 183–87).*
  (d) Consider the presence of eyespots on the wings:
    (i) Specimen is boldly patterned, including eyespots: *see True Brushfoots such as the two Emperors (pp. 370–73), Common Buckeye (pp. 342–43), Painted Lady (pp. 356–57), or American Lady (pp. 358–59).*
    (ii) Specimen has a combination of eyespots and somber coloring (grays and browns): *see brushfoots/Satyrs and Wood-Nymphs (pp. 48–49, pp. 374–97).*

(3) Small wing size (less than 1.5 inches)
  (a) Tails present: consider tail size:

(i) Hindwing tails are thin and threadlike: *see gossamer-wings/Hairstreaks (p. 43, pp. 256–89); check also the Eastern Tailed-Blue (pp. 298–99).*

(ii) Hindwing tails are thick and project well past the end of the body: *see spread-wing skippers/tailed-skippers (p. 39, pp. 56–61).*

(b) No tails: consider wing color:

   (i) Dorsal wings are blue: *see gossamer-wings/Blues (p. 43, pp. 290–309); also check the White M Hairstreak (pp. 287–89).*

   (ii) Wings are a combination of yellow/off-white and/or orange with black markings and margins: *see smaller sulphurs such as the Clouded Sulphur (pp. 225–26), Orange Sulphur (pp. 227–29), Barred Yellow (pp. 242–44), Little Yellow (pp. 247–49), Sleepy Orange (pp. 250–51), or Dainty Sulphur (pp. 252–53).*

   (iii) Dorsal wings are primarily bright white: consider color of ventral wings:

      (1) Ventral hindwing has a heavy, greenish mottled pattern: *see Falcate Orangetip (pp. 222–24).*

      (2) Ventral wings are primarily white/yellowish white or lightly mottled with yellow or gray: *see whites and sulphurs/Checkered White and Cabbage White (pp. 41–42, pp. 212–16).*

   (iv) Dorsal wings appear orange and/or brown with various darker markings: consider the following:

      (1) Long "beak" present: *probably an American Snout (pp. 312–13), easy to identify once seen at rest.*

      (2) Flies with a leisurely flap-and-glide flight pattern: *see brushfoots/crescents and checkerspots (pp. 45–46, pp. 328–41).*

      (3) Flies very low to the ground in a fluttery, mothlike pattern: *see Little Metalmark (pp. 310–11).*

      (4) Flies in a frenetic, hard-to-follow manner: *see Harvester (pp. 254–55).*

(c) Thick, stocky body, short wings, and hooked or angled antennae: start with the skipper accounts and consider wing profile when perched:

   (i) Wings are held flat when perched: *go to Spread-wing Skippers (p. 39, pp. 54–94):*

      (1) Dorsal wings are dark brown to black with clear or white spots, particularly on the forewing: *see cloudywings (pp. 64–69), duskywings (pp. 72–87), and Common Sootywing (pp. 93–94).*

      (2) Dorsal wings are white/gray and black-mottled pattern on all four wings: *see checkered-skippers (pp. 88–92).*

spread-wing profile

jet-fighter profile

(ii) Specimen is small with triangular wings held closed over body in a "jet-fighter" profile when perched: *go to Grass Skippers (pp. 39–40, pp. 95–182)*:

  (1) Specimen is very small and orange/tan colored: *see Least Skipper (pp. 102–3) and Southern Skipperling (p. 104–5)*.

  (2) Specimen is very small and dark colored with varying degrees of white markings: *see roadside-skippers (pp. 159–68)*.

  (3) Primary wing coloring is dark/chocolate brown:

    (a) Speciman has varying degrees of white spots and/or frosting: *see Clouded Skipper (pp. 100–101), Dun Skipper (pp. 154–55), Dusted Skipper (pp. 156–58), and Twin-spot Skipper (pp. 171–73)*.

    (b) Specimen is found along Gulf Coast: *see Neamathla Skipper (pp. 97–99), Salt Marsh Skipper (pp. 177–78), Obscure Skipper (pp. 179–80), and Ocola Skipper (pp. 181–82)*.

  (4) Specimen is primarily yellow/tan: take note of location:

    (a) Specimen is found in an open habitat such as fields, open roadsides, parks, sports complexes: *see Fiery Skipper (pp. 106–7), Sachem Skipper (pp. 112–13), Tawny-edged Skipper (pp. 114–15), Crossline Skipper (pp. 116–17), Whirlabout (pp. 118–19), and Southern Broken-Dash (pp. 120–21)*.

    (b) Specimen is found around water such as marsh or swamp, or in coastal prairie along the Gulf Coast: *see Delaware Skipper (pp. 130–32), Broad-winged Skipper (pp. 138–40), Aaron's Skipper (pp. 143–44), Palatka Skipper (pp. 145–46), Dion Skipper (pp. 147–49), Bay Skipper (pp. 150–51), and Dukes' Skipper (pp. 152–53)*.

BUTTERFLIES AND SKIPPERS BY FAMILY

Skippers (Hesperiidae)
SPREAD-WING SKIPPERS (PYRGINAE)

There are twenty-six members of the spread-wing family along the US East Coast and nineteen in LA. This family includes the longtailed skippers, cloudy-wings, duskywings, and checkered-skippers. They have large heads with antennae that end both at an angle and in a sharp point. Their bodies are short and stubby. The colors are mostly dark brown to blackish-brown. Some, like the checkered-skippers, are mottled; others possess translucent spots on the forewings. At rest, many hold their wings in an open, thus "spread-wing" fashion; however, some fold their wings upward. Many of the males have a costal fold along the leading edge of their forewings. This subfamily is fond of flowers, but can also be regularly found around mud. Most males perch, but some will patrol.

Dorantes Longtail
(dorsal)

Funereal
Duskywing
(dorsal)

GRASS SKIPPERS (HESPERIINAE)

Forty-one species of grass skipper have been recorded in LA, some of which have not been seen within the state in many years. Most are grass feeders. These are fast-flying butterflies with typically shorter, pointed, or triangular wings and stout, thick bodies. Their heads are large with hooked ends on the antennae. The colors are mostly shades of yellow, tan, and brown with a variety of dark, light, or translucent spots or other markings. Many of the males have a dark line or patch near the middle of the dorsal forewing. At rest, they close their wings over their body, but many will also bask with their wings in the jet-fighter profile. The adults have six working legs.

Yehl Skipper (dorsal)

Dusted Skipper (ventral)

GIANT SKIPPERS (MEGATHYMINAE)

Giant Skippers are exclusively a New World family with most occurring in the US Southwest and Mexico. There are two members of the family recorded in LA, but one has not been seen much of late. Both species within the state use yuccas as their food plant. These are the largest skippers, with wingspans of two to three inches. Their bodies are typically described as cigar-shaped. Their head is narrower than the thorax, and their antennae are clubbed at the end. They are extremely swift fliers that rarely perch.

Yucca Giant Skipper (dorsal)

Butterflies (Papilionoidea)
SWALLOWTAILS (PAPILIONIDAE)

These large, showy butterflies are present around the world. There are over thirty different species in North America and eight in LA (one being an extremely rare stray). Within this family, there are three subfamilies within the state—the pipevine-feeding swallowtails, the kite swallowtails, and the fluted swallowtails—of which there are two, one, and five species, respectively. Louisiana's largest butterflies, they are characterized by having tails on the hindwings, thus yielding the name "swallowtail." In addition to their large wings and tailed hindwings, they possess six working legs, large eyes, and large and stout bodies. Their coloring is mostly a dark/black background with yellow and/or blue markings, but also include yellow-striped and white-striped patterns. While feeding, some flutter their wings while others hold them open. At rest, most hold their wings in an open fashion.

Giant Swallowtail
(a Fluted Swallowtail, dorsal)

Zebra Swallowtail (a Kite Swallowtail, dorsal)

WHITES AND SULPHURS (PIERIDAE)

There are about sixty species across North America, with five whites (one an extremely rare stray) and thirteen sulphurs (three recorded as strays) in LA. The whites are placed in the Pierinae subfamily and possess rounded white wings with black markings and/or margins. The sulphurs fall into the Coliadinae subfamily and include species with rounded yellow, orange, or white (the alba females) wings. Smaller than swallowtails, whites and sulphurs are small- to medium-sized butterflies, although a couple of the sulphurs can approach the size of male Black Swallowtails. There is both seasonal variation (with darker wing scaling in the spring and late fall) and sexual differences (with the females showing more markings) in coloration in both groups. The whites typically use crucifers (mustards) as food plants, while most of the sulphurs in LA use legumes. Like the swallowtails, the adults in this family have six working legs and large, robust bodies. Most are strong fliers. At rest, they usually hold their wings in a closed position over their bodies but will open those wings to bask in the light of the sun.

Checkered White (dorsal)

Great Southern White

Cloudless Sulphur (ventral)

Sleepy Orange (ventral)

GOSSAMER-WING BUTTERFLIES (LYCAENIDAE)
Harvesters (*Subfamily Miletinae*)
Harvesters are primarily found in Africa and Asia. There is only one species in North America, including LA. The caterpillars of this family are also the only ones in North America that are carnivorous, feeding on aphids. The adults feed on honey-dew secretions produced by the same aphids upon which its larvae feed. Like hairstreaks, these are small butterflies with a swift and erratic flight. They are orange-brown and yellow with marking both dorsally and ventrally. When at rest, harvesters hold their wings closed over their bodies like hairstreaks.

Harvester (ventral)

Hairstreaks (*Subfamily Theclinae*) and Blues (*Subfamily Polyommatinae*)

There are over one hundred species of hairstreaks and blues in North America north of Mexico, with fourteen hairstreaks (one stray) and nine blues (three occasional strays) in Louisiana. Hairstreaks are in the Theclinae subfamily, which also includes the elfins. The blues are in the Polyommatinae subfamily. Both hairstreaks and blues are small, delicate butterflies. Certain species are tended by ants as part of a symbiotic relationship. Adult lycaenids rest with their wings held in a closed position over their back. In most instances, the males have only four working legs (the front two legs are not functional), while the females have six. The eyes are large, but the bodies are thin and delicate. In LA, almost all of the hairstreaks have tails. They possess intricate patterns on their ventral wings (stripes, spots, lines, patches of red or blue color), and show a variety of colorations dorsally (black, brown, deep blue, purple, and gray). Their flight is fast and can be hard to follow. The blues are mostly smaller than the hairstreaks and have bright blue dorsal wings with silver or gray ventral wings that possess small black or brown markings. Their flight is typically close to the ground and less swift.

Great Purple Hairstreak (ventral)

Spring Azure (ventral)

METALMARKS (RIODINIDAE)

There are over a thousand species of metalmark in the American tropics; only around twenty-five have been recorded north of Mexico, and only one in LA. These are small, delicate butterflies. Most in North America are yellow and/or brown with intricate patterns both dorsally and ventrally that include flecks of metallic markings, which give them their name. Their flight is shallow and low to the ground. They perch, often on the underside of leaves, with their wings in an open position. Like lycaenids, the males have only four functioning legs while the females have six.

Little Metalmark (dorsal)

BRUSHFOOTED BUTTERFLIES (NYMPHALIDAE)

This family goes by the common name of brushfoots because the forelegs of both the males and females are significantly reduced, nonfunctional, and possess brushlike scales. There are over 160 species of this family in North America north of Mexico, 22 found in LA. The members of this greatly diverse family come in a variety of sizes, colors, and designs. Some have hindwing tails; others possess angular, ragged wing margins. Typically, its members have distinct knobs on the end of long antennae and are strong fliers with multiple markings and colors on the wings. They are territorial, and many are long-lived. They may be found around flowers, sap, mud, fermenting fruit, dung, or carrion. Some perch with their wings closed over their body, while others leave their wings open. Many will bask with their wings half-open. The adult bodies are usually large and stout. It is not always easy (and many times impossible) to determine if the individual butterfly being inspected has four or six working legs.

This family includes the following subfamilies: True Brushfoots (Nymphalinae), Greater Fritillaries (Heliconiinae; subgenus *Speyeria*), Crescents and Checkerspots (Melitaeini), Admirals (Limenitidinae), Leafwings (Charaxinae), Emperors (Apaturinae), and four subfamilies described in more detail below: Snouts (Libytheinae), Milkweeds (Danaidae), Longwings (Heliconiinae), and Satyrs and Wood-Nymphs (Satyrinae).

Gorgone Checkerspot (dorsal)

Red Admiral (dorsal)

Eastern Comma (dorsal)

Tawny Emperor (ventral)

Viceroy (dorsal)

Goatweed Leafwing (dorsal)

Snouts (*Libytheinae*)

There are ten snout species worldwide, two of which occur in North America. One is present in LA. This is considered one of the more ancient of the butterfly families. The name derives from the long beaklike projection, known as palpi on the head of the adults. This projection looks like a long snout. This is a smaller butterfly, larger than hairstreaks but smaller than most brushfoots. Its coloration is cryptic, browns and grays ventrally, such that when it perches with its wings closed and folded over its body, it looks like a dead leaf with the

American Snout (ventral)

snout as the leaf's stem. Dorsally, it is a combination of brown and orange. Its flight is swift and hard to follow. Like the gossamer-wings, the males have four functioning legs while the females have six.

Milkweeds (*Danaidae*)

These are larger butterflies. Last-season female Monarchs are some of the largest butterflies on the wing in the state. Their color is distinctive, orange and brown with black markings and white spots, believed by many to reflect universally recognized warning colors to the avian world and resulting from the toxin in the milkweed plants upon which they feed. These butterflies have a strong soaring, gliding flight. Like other nymphalids, both sexes have only four walking legs.

Monarch (ventral)

Longwings (*Heliconiinae*)

The longwings are primarily tropical butterflies found in Mexico, Central America, and South America; however, four species (one doubtful) have been recorded in LA. These are moderately large butterflies with long, narrow wings (the forewings are twice as long as they are wide). Their bodies are long and slender, and their antennae are almost as long as their bodies. The forelegs are feeble in both sexes. Their wing beats are typically shallow. The males patrol, and both sexes are extremely attracted to flowers. All of the longwings found in LA feed on passion vines and are believed to be toxic.

Gulf Fritillary (ventral)

Satyrs and Wood-Nymphs (*Satyrinae*)

These butterflies are primarily small and midsized brown and gray butterflies with multiple eyespots dorsally (on some) and ventrally (on all). There are about fifty species in North America with nine (one very rare) in LA. Like the other brushfoots, Satyrs and Wood-Nymphs have reduced forelegs in both sexes. Primarily forest and forest-edge species, none should be expected to be found at flowers. Their flight is distinctively bouncy and low, in and just above the grass. They will perch on the ground or in the grass with their wings folded up over their back, occasionally opening the wings to about 45 degrees while basking.

Northern Pearly-eye
(ventral)

Georgia Satyr
(ventral)

SPECIES ACCOUNTS

For each species entry, I include the following information:

Photographs: I use primarily photographs of live specimens, presenting dorsal and ventral views. I include male and female photos when the species is dimorphic. Date and location of each photo are given in the Notes on Photographs at the end of the book.

Description and Behavior: While I have primarily relied upon photographs to assist the reader in identifying a particular species, many useful guides include written descriptions of wing markings. For example, *How to Know the Butterflies* (1961), by Paul and Anne Ehrlich, relies almost exclusively on written descriptions. As a supplement to the photographs, I have also noted features that distinguish the species from others similar in appearance. I have also identified traits and characteristics that I and others have observed about the species. I have indicated which species have been classified as possessing SGCN (Species of Greatest Conservation Need) status in LA. There are three tiers—I, II, and III—in descending order of priority, describing those species' need for conservation action to sustain or restore their populations within the state. These species are further graded as S1 (critically imperiled), S2 (imperiled), S3 (vulnerable), and SU (unranked due to lack of information).

My Records: I cite regions of the state, parishes, and sometimes specific locations where I have seen the species, as well as the months when I saw it.

Distribution and Abundance: I draw on records from various sources to describe where and when the species has been found over time. Included here are results from the Lepidopterists' Society Season Summaries and NABA counts, published articles, and individuals' observations.

Host Plants: I identify by common name the host plant(s) that the species most probably feeds on in LA. In many instances, I use data from TX, AL, or the South in general. In some circumstances, I include a specific butterfly-plant relationship that I feel the weight of evidence has proven to exist in LA. I have addressed only minimally larval food plants. Appendix B gives the scientific names of the plants mentioned here.

Distribution Map: Using a map of Louisiana divided into parishes, I indicate with a red-orange dot every parish where the species has been recorded.

Note: I do not address in detail the various subspecies that might inhabit LA. Unlike the western section of the US, few of LA's species have more than one subspecies within the state or present a subspecies here that would be different from the rest of the eastern portion of the country. I have discussed subspecies for a few species (for example, Queens, Viceroys, Texan Crescents, and Aaron's Skippers) that are roughly on the border between the ranges of eastern and western (or northern and southern) subspecies. Jonathan Pelham's *Catalogue of the Butterflies of the United States and Canada* (2008) is an excellent source of information on all subspecies recognized in the United States.

SKIPPERS

(Superfamily Hesperioidea; Family Hesperiidae)

Silver-spotted Skipper (*Epargyreus clarus*)

ventral

dorsal

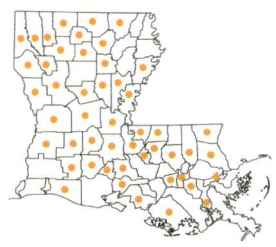

Description and Behavior: With a large silver spot on its lower ventral wing, this skipper is easily recognized. It is a swift flier and an avid flower visitor. Israel (1981) recorded it at seventeen different species of flowers and at mud in the Tunica Hills region.

My Records: My records are from Apr until Oct. I believe there are at least three broods (as did Israel), Apr and May, July to Aug, and then late fall, Sept to Oct. I regularly see it within the city of Lafayette and believe it uses evergreen wisteria as its host plant. It is common at Thistlethwaite WMA, and in Aug and Sept can be found visiting blooming ironweed. At Indian Creek Rec Area and Avery Island (at the rookery), this skipper is fond of blooming buttonbush, where it can be so engrossed in feeding that I believe I could pick them off with my fingers. It was present in June and July at Bayou Cocodrie NWR, Sicily Island Hills WMA, and Copenhagen Hills.

Distribution and Abundance: It has been regularly reported from around the state. In CAD, it has been seen from early Mar to Oct, with a single sighting in Nov, and elsewhere in north LA on numerous other occasions. Raw data from the Cajun Prairie survey reflected sightings in five parishes. It appears to be encountered less frequently in southwest LA. It is common in both of the Felicianas, flying from Mar until Nov. In the River Parishes, three were found in Vernon Brou's light traps in SJB. In south LA, it has been recorded in July on Grand Isle. It has been reported on numerous NABA counts over the years, forty-three individual counts in all. These include the Alexandria, Catahoula, Pearl River, New Orleans, Shreveport, Global Wildlife, Folsom, West STA, and Bonnet Carré counts. On the 2010 Shreveport count (Sept), twenty-two were reported, and on the 2010 Global Wildlife count, thirty-four were recorded.

Host Plants: It is reported to use red clover, vetch, sweet acacia, hyacinth bean and wisteria, kudzu, members of the bean family (bastard indigo, hog peanut, butterfly pea, milk peas, and tick-clovers), and members of the locust-tree family (black, honey water) as its larval food plant.

White-striped Longtail *(Chioides catillus)*

ventral

dorsal

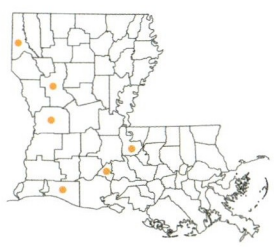

Description and Behavior: This skipper is an occasional visitor from south TX into the TX Gulf Coast. The easily visible white stripe on its lower ventral wing is distinctive. Its flight is similar to the two other long-tailed skippers found within LA, swift and bouncing, and it is a frequent flower visitor.

My Records: I've seen this skipper only once in LA, in 2007, at the Kisatchie unit in NAT. It was in a hillside pitcher-plant bog, near a freshwater spring on Forest Service Road 280.

Distribution and Abundance: Past records are exclusively in the western half of LA. The first sighting was in early Sept 2006, in VER, along Hwy 111. It has shown up twice in the NAT unit of Kisatchie NF, the second time in mid-Aug 2012 (see My Records for the first). In north LA, it has been found twice in the Shreveport area in Sept 2006, and in mid-Nov 2012. In southwest LA, one was seen at Peveto Woods in CAM in early Apr 2008, and again in late Oct 2012. Other recent sightings (and the most eastern) include three during one day and five the next within Sherburne WMA in late Sept 2013. The earliest (and only spring) sighting was in LAFA at the Acadiana Nature Station on Feb 19, 2017.

Host Plant: In the Texas hill country, it uses legumes.

SKIPPERS

Long-tailed Skipper (*Urbanus proteus*)

ventral

dorsal

Description and Behavior: This is the most common tailed skipper in LA. The three tailed skippers recorded in LA are easy to distinguish. The white stripe on the White-striped Longtail is distinctive and unmistakable. The Long-tailed Skipper has a green sheen dorsally on its hindwings as well as longer tails. The Dorantes Longtail has neither the white stripe nor the green coloring.

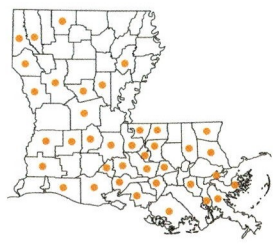

After a mild winter, this longtail can be seen by midsummer; otherwise, it begins to arrive in early fall, and is often one of the last butterflies on the wing. It favors open sunny places, is an avid flower visitor, and is easily drawn to a city garden if lantana is present. More times than not, late-fall individuals are lacking one or both tails, and I have on several occasions been stumped as to what I was seeing, only to realize it was a completely tailless Long-tailed Skipper.

My Records: My records, mostly in the Acadiana region, are from Apr into early Nov. I've also seen it from late Sept to Nov in CAM. In the eastern portion of the state, I've seen it very early at Sandy Hollow WMA (twice in Mar) and at Abita Creek Flatwoods Preserve (late Sept). The winters of 2009–10 and 2010–11 had long stretches of freezing weather. As a result, this skipper was scarce in 2011. I did not see any in 2012 until Sept, at which time it became common, flying into Nov.

Distribution and Abundance: It has been recorded from Feb to early Dec. Records from north LA include several specimens within the State Collection at LSU from the Winnsboro area in FRA, collected in a soybean field. It is common in CAD from Aug into Nov. In the central LA region, it has been recorded in the NAT and RAP units of Kisatchie in Aug and Sept, respectively. It has been recorded in BEAU in Oct. It has been recorded only in the fall in WFE, where it appears to be rare, and all months (except Feb, May, and Aug) at Asphodel Plantation, where it is more common. In the south, there are records from Grand Isle, Galliano, Golden Meadow, Port Fourchon, and Larose, flying from July to Dec (with one sighting in Mar).

This skipper has been reported on several LA NABA counts: Catahoula (Sept 2010 and Oct 2012 and 2013), New Orleans (June, seven counts since 1995), Bonnet Carré (Aug 1998, 1999, 2000, and 2005), Folsom (Oct 2003 and 2004 and Sept 2006), Grand Isle (Aug 2004), Cameron (July 2000 and Aug 2003), and Global Wildlife (July 1999 and 2000).

Host Plants: Generally, this skipper uses members of the locust-tree family (black, honey water). Other identified plants include certain species of the bean and pea family such as hyacinth bean, false indigo, green beans, and potato bean and include members of the wisteria family.

SKIPPERS

Dorantes Longtail *(Urbanus dorantes)*

ventral

dorsal

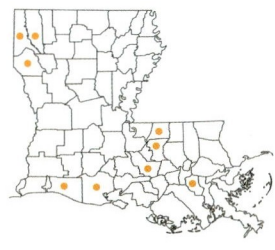

Description and Behavior: This skipper is a regular member of the butterfly fauna in south FL and south TX. It is less distinctive than its two cousins, with some off-white spots dorsally and dark splotches ventrally. There does not appear to be any discernible pattern to its distribution from this skipper's migration into the state during the fall months.

My Records: I have seen it in CAM in late fall. In fact, in 2010 and 2012, I saw several over the last three weeks of Oct. I found one at Freshwater Bayou Lock Rec Area in late Oct 2014.

Distribution and Abundance: The first report of this skipper was in EBR in Sept. In north LA, three individuals have been recorded in Sept and Oct in CAD. It has also been recorded at Red River NWR (Oct) and the Stonewall area of DES (Sept). In southwest LA, there have been additional sightings at Intracoastal City and Peveto Woods (Oct). In the middle of the state, there was a Nov sighting in IBV. In the FL Parishes, there have been two sightings at Asphodel Plantation in Oct and Nov, and several at Taft.

Host Plants: In south FL and south TX, it is reported to use members of the bean family, butterfly pea, beggar's ticks, and pigeon-wings.

SKIPPERS

Hoary Edge (*Achalarus lyciades*)

ventral

dorsal

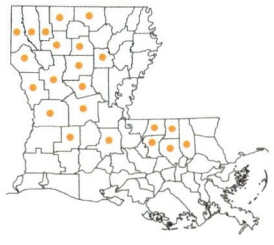

Description and Behavior: This distinctive spread-winged skipper has a series of connected yellow spots on the upper wings and a bright white splotch on the lower ventral wing. It is a larger skipper, a bit smaller than a Silver-spotted Skipper. The females are typically larger. It is a swift flier; however, when taking nectar it can be easily approached. Blooming buttonbush is a magnet for this skipper. I have to admit that, at times, I have thought older individuals to be Gold-banded Skippers, particularly when the white splotches on the lower, ventral wings are worn and hard to distinguish.

My Records: My records are from Apr to Aug, more common in the spring. I have found it to be common in the piney woods of central LA, taking nectar from numerous wildflowers such as false garlic, blackberry, and dandelions. On occasion, it has appeared in good numbers at the GRA unit of Kisatchie NF, both at the Catahoula Butterfly Garden and along Stuart Lake Road. It has also shown up regularly in the NAT, VER, and RAP units. I also found it at Copenhagen Hills in June and July. The farthest south I've seen it has been at Thistlethwaite WMA, where it is rare. It has not been reported on any of the NABA counts in the southern or eastern portion of the state, so far as I have been able to determine.

Distribution and Abundance: Old records were only from northwest LA in June and Aug. In CAD, it has been found from early Apr to late Sept. Other north LA sightings include Driskill Mountain (Apr), Sparta Quad (July), Bodcau WMA, Red River NWR, Cypress Lake (Apr–May and July–Sept), and Caney Lake Rec Area (July). In the central LA region, there are additional records from ALL (Apr) and SAB (May). It has been reported in both of the Felicianas as uncommon to common over two broods flying from early June to early Sept. In the FL Parishes, it has been found near Greensburg (Apr) and Fluker (July and Aug).

Host Plants: In its southern range, it reportedly uses clovers, tick trefoils (the seeds of which my father calls "stick tights"), and false indigo.

Northern Cloudywing *(Thorybes pylades)*

ventral

dorsal

Description and Behavior: This dark, nominally marked cloudywing can be very common in the spring. In mid- to late Mar, it is often the dominant cloudywing, but by early Apr, its cousin, the Southern Cloudywing, is more common. To add to the confusion, its other cousin, the Confused Cloudywing, can also be found flying in the same area. All

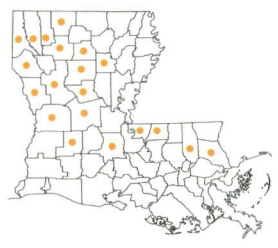

three cloudywings fly low to the ground. All love to visit many kinds of flowers, at which time they are approached most easily; however, even at flowers they remain wary. Cloudywing flight is swift, but if not disturbed, they will quickly stop at another flower. The spring brood is small, but as the season progresses, this skipper (as well as the Southern Cloudywing) increases in size. During the summer, the females can be as large as a Hoary Edge.

My Records: My records include Mar through May and then again July and Aug, suggesting two or possibly three broods. Israel (1981) reported three broods. There have been days at the GRA unit of Kisatchie NF when this cloudywing has been the most common butterfly present, seen along virtually every road, stopped at numerous kinds of wildflowers. One time in late Apr 2008, it seemed that every flower along the Stuart Lake Road in that unit was occupied by this skipper. Open fields in the NAT unit of Kisatchie are also good places to find it. I found it at Cooter's Bog in early July and at Copenhagen Hills in late July.

Distribution and Abundance: Often common when found, the Northern Cloudywing has been only sporadically recorded in LA. First reported in 1965, all of the older records were from the pine-forest region of central LA. In the years since, it has also shown up in both north LA and the FL Parishes. In north LA, sightings have been in late Mar, Apr to May, and then sporadic sightings in mid-July, Aug, and Sept. Some specific records include Bodcau WMA and Cypress Lake (Apr–Aug), Sparta Quad (July), and Driskill Mountain (Apr) and Kisatchie NF in central LA (Apr). It has been recorded as rare in EFE, from Apr and again during July to Aug. The only NABA counts to record it have been the Shreveport and Global Wildlife counts.

Host Plants: In southeast TX, it is reported to use beggar's ticks, bush clovers, and other legumes.

Southern Cloudywing (*Thorybes bathyllus*)

ventral

dorsal

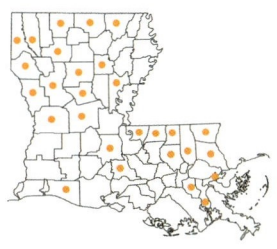

Description and Behavior: A question often asked is how to differentiate this cloudywing from its cousins. First, the Northern flies earlier than the other two, so anything in early Mar is probably that skipper. The Southern has large white spots on the upper dorsal and ventral wings. The Northern has very small, if any, white spots. The Confused is somewhere in between. The Southern and Confused have white circles around their eyes and a white face; the Northern does not. There is a white patch on the top of the antenna bend on Southerns, but not on the other two. The upper/outermost line of white spots on the dorsal upper wing (usually consisting of four dashes) is straight on the Southern but offset on the Confused. Per Gatrelle (2001) on this species, "The four tiny hyaline spots at the apex of the forewing are ALWAYS in a straight line." On the Confused, "the line of tiny spots at the apex ALWAYS has the last spot noticeably disjunct toward the apex." Finally, the Southern's wings are typically fringed in white, almost silver. On the Confused, that fringe is more tan to light brown while the Northern has little fringe, if any at all.

My Records: Of the three cloudywings found in LA, this one is consistently the more common over the course of an entire year. My records are from both the Acadiana and central LA regions, from Apr to May and then July to Aug, representing at least two separate broods. A sure place to find it is in the NAT unit of Kisatchie NF in Apr in open fields with nectar sources. The latest date I have found it was at the end of Aug, at Cooter's Bog. Like the Northern Cloudywing, it flies swift and low, stopping often at many kinds of wildflowers.

Distribution and Abundance: Older records from LA were from late Feb through early Sept, found primarily in the FL Parishes. It has been recorded on occasion in north LA (Apr, May, and July) at locations such as Driskill Mountain (Apr) and Bodcau WMA (July). At Asphodel Plantation it is rare in June, Aug, and Sept. This cloudywing has shown up on LA's NABA counts far more often than either of its cousins. These include the Folsom (2004), Barataria (1998), Global Wildlife (multiple years), Copenhagen (2014), Cameron (1996), Pearl River (1981, 1994), Bonnet Carré (1996), New Orleans (2006 with 36 reported), West STA (2000), and Shreveport (2008) counts.

Host Plants: On the East Coast, various legumes are reported as used.

Confused Cloudywing *(Thorybes confusis)*

ventral

dorsal

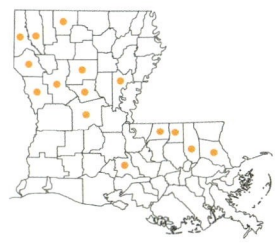

Description and Behavior: Lambremont (1954) was the first to report this cloudywing, but used the common name of "Northern Cloudywing." He observed that it was found "occasionally" here during the summer in "open fields and sunny places visiting flowers, often clover." When all available records are considered, it appears there are two broods; however, there may be a third brood present between the spring and fall broods. The first would be in Apr with another in July and Aug. Its habits and behavior match those of its two cousins. I've found it to be the least common of the three cloudywings in LA.

My Records: I have seen it multiple times in the NAT unit of Kisatchie NF, mostly in Apr, found flying in the same open fields where its two cousins can be present. It was also at Dove Field in July in the VER unit.

Distribution and Abundance: Older records for this uncommon cloudywing were from north LA and the FL Parishes, June through Aug. It is uncommon in north LA (including Bodcau WMA), with sightings in late July into Aug. Recent records from central LA include the GRA, NAT, and VER units of Kisatchie from May through July. A male specimen of this species was photographed by Clark in LAS in mid-June of 2017 (a late record not reflected in the distribution map). It is rare at Asphodel Plantation from May to July. It was reported once on the Global Wildlife count (July 1994) and once on the Pearl River count (1976).

Host Plants: In the east, it is believed to use bush-clover and other members of the pea family, but various legumes, for example tick trefoils, and wild indigo are also suspected.

Hayhurst's Scallopwing (*Staphylus hayhurstii*)

ventral

dorsal

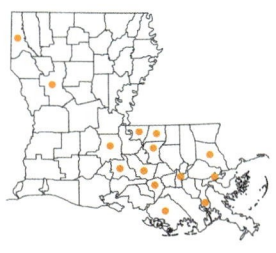

Description and Behavior: These are small, dark skippers. They fly low, don't tend to move around a lot, and can be easily overlooked. They could possibly be confused with the Common Sootywing as their appearance is similar; however, their flight pattern is different. The flight pattern of the Hayhurst's Scallopwing is typically slower and more fluttery. In contrast, the Common Sootywing's flight is swift and darting like a skipperling.

My Records: I've seen it at Indian Bayou WMA several times, primarily in the heat of June and July in oak hardwoods inside the levee road along the Atchafalaya River. Typically, they were flying in the low grass along several four-wheeler trails. Recorded several times as part of the Indian Bayou count, the high was forty-six in June 2015, at multiple locations over the WMA. Once, at Indian Bayou WMA in early July, in the midst of an extended dry spell, I found it at mud in a creek bed that still had a little water.

Distribution and Abundance: Lambremont's oldest record was from JEF, in Mar in a "dense oak-hardwood stand on the front lands of the Mississippi River" (1954). Records from north LA (CAD, Sept) suggest it is rare there. I found only one record from the central LA region, Oakland Plantation in Apr 2005. In the Acadiana region, it was reported at BugStock (a weekend-long search for and study of arthropods on private property near the town of Washington) in SLA in 2007 (Sept) and 2010 (June). There are a few records from the center of the state (June to Sept). Israel (1981) reported it in both of the Felicianas, seen in Apr and Sept, and described it as extremely rare during his survey. In the FL Parishes, multiple specimens have been taken in Apr, Aug, and Sept. It has shown up one time on the New Orleans count (July 1992), and once on the Pearl River count (1993). In southernmost LA, it has only been recorded in TER (May).

Host Plants: In the Texas hill country, it is reported to use pigweed and lamb's quarters. In south FL, it uses alligator weed.

Sleepy Duskywing (*Erynnis brizo*)

ventral

dorsal

dorsal male

ventral

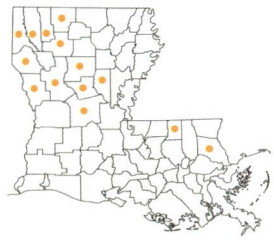

Description and Behavior: In early Mar, the Sleepy Duskywing is one of only a couple of spread-wings flying. At that time, there is not a lot of green undergrowth in the pine and scrub-oak habitat this duskywing favors, and its black color is easily seen as it moves about. When Horace's and Juvenal's duskywings begin to fly in late Mar, neither is as dark black as the Sleepy. Also, when basking, as all three are prone to do in the spring, the Sleepy's diagnostic gray bands and different spot pattern can be seen. The males tend to land often, such as basking on dirt roads, but are easily disturbed and often blown away in the ever-present Mar wind.

My Records: My records are all in the central LA region. I primarily find it in Mar, flying in the same areas as the several Frosted Elfin colonies in NAT, RAP, and GRA. There is a large open field next to the Catahoula Butterfly Garden in GRA, and this duskywing can be found flying in that field. In the NAT unit of Kisatchie NF, the males patrol in the understory of the pine flats and don't seem to venture out into the open as they do in GRA. In the RAP unit, they fly along a power-line cut that crosses Castor Plunge Road.

Distribution and Abundance: This duskywing's records are primarily west of the Mississippi River. It was first reported by Strickland (1972), who found it in LAS in Mar and Apr. His comment was, "It is generally associated with pine-scrub oak hills." Records from north LA include Driskill Mountain in Apr; Bodcau WMA in late Mar; Walter Jacobs Nature Park in mid-Mar, near Hunter (DES); Kisatchie NF in WEB in Apr; and Caney Lake Rec Area, also in Apr. The only confirmed records I found from east of the Mississippi River were by Strickland, who found it in near Greensburg. Six were reported during the 1997 Pearl River count.

Host Plant: Scrub oaks.

Juvenal's Duskywing (*Erynnis juvenalis*)

ventral

dorsal
female

dorsal
male

Description and Behavior: In the spring, this and Horace's Duskywing fly to-gether and exhibit similar appearance and behavior. They are best distinguished by looking for two white spots along the leading edge of the ventral hindwing. If present, it is Juvenal's Duskywing. Israel (1981) suggested there was only one brood, in the spring; however, multiple records from June and July suggest a second brood, and the Sept records may reflect a third partial brood.

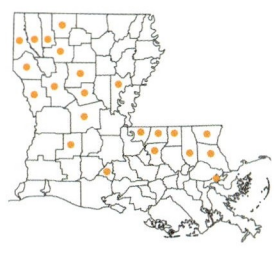

My Records: For the most part, my records from NAT, GRA, and RAP reflect a spring flight in Mar and Apr. I also found it at Hutchinson WMA in Mar and in WFE in July.

Distribution and Abundance: This duskywing is common to abundant in CAD (Eddie Jones Park, Walter Jacobs Nature Park) from late Feb to early Apr. Other north LA records are only from Mar and Apr. In the central LA region, there are older records from Feb, Mar, and early Apr, and one recent record from Acadiana (Acadiana Park, Apr). The remainder of records for this skipper are from the FL Parishes (Mar and Sept). It is reported to be common in both EFE and WFE, with records from late Feb into June. It has been reported three times on the Pearl River count (typically conducted in June).

Host Plant: This duskywing is reported to use oaks. For example, Ross and Lambremont (1963) reported that twenty-five larvae and pupae were taken on water oak in Sept in EBR.

Horace's Duskywing *(Erynnis horatius)*

ventral

dorsal female

dorsal male

Description and Behavior: I believe this to be the most common duskywing in LA, flying in all of the warm months in multiple habitats and locations. It is fond of numerous kinds of low-growing flowers in open fields and along roads and trails, and it is observed most easily while feeding at those flowers. Israel (1981) reported seeing it at seventeen different species of flowers as well as mud and dung. I once counted fifteen at a wet spot in a parking lot at Bayou Cocodrie NWR. It seems to prefer hot, sunny spots in and along dense forests. All of LA's duskywings will perch on spots of bare earth, but the Horace's (as well as the

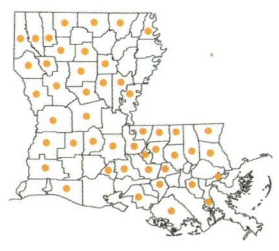

Juvenal's) will also perch on the outer leaves of its primary food plant, oak trees. Neither Sleepy nor Wild Indigo Duskywings perch in trees. I also have noticed that this skipper tends to fly higher (but not always) than Sleepy or Wild Indigo Duskywings. Its flight is fast and evasive. I don't find that the males patrol in any discernible pattern; rather, they perch to await a passing female.

The females are not as dark as the males and have more visible dorsal wing patterns. In the spring, the dorsal patterns on the males are also more visible. There are at least three, if not more, broods in early spring, midsummer, and fall. The fall broods are extremely dark and difficult to differentiate from Zarucco Duskywings, which, in most locations, will be the only other duskywing flying after July. (Wild Indigos will fly that late but only in areas with wild indigo.)

My Records: I've found it from Mar to Sept throughout the western portions of LA. It is extremely common at Thistlethwaite WMA and in the NAT, VER, and GRA units of Kisatchie NF. It is also common at Mary Ann Brown Preserve.

Distribution and Abundance: In north LA, it is common to abundant in CAD (and other parishes) from late Feb to early Oct. In the central LA region, it has been recorded from Feb to Dec. (The Feb records were actually from CALD with caterpillars found on two kinds of oak.) In the Felicianas, it is common to abundant, with sightings from Feb to Nov. It has appeared regularly on several NABA counts from around the state. In north LA, thirty-six were reported on the 2010 Shreveport fall count. In southwest LA, it has been recorded on the Cameron count. In the FL Parishes, seventy-five were reported from Global Wildlife in 2000 with another twenty-five in 2002. It has shown up on twenty-one separate Pearl River counts and on twelve New Orleans counts.

Host Plant: Like its cousin species, *juvenalis*, it uses oaks.

Mottled Duskywing *(Erynnis martialis)*

dorsal

dorsal

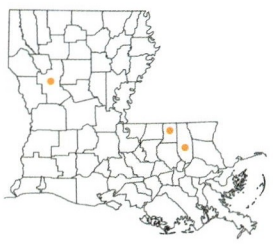

Description and Behavior: In the southern US, this duskywing's range includes the FL Panhandle, across the Gulf States, and into central TX. It is reported as rare and declining across most of its eastern range. Schweitzer, Minno, and Wagner (2011) suggested that the reasons for this decline include increased deer populations grazing on its food plant as well as excessive prescribed burning. Throughout most of that range, the Mottled Duskywing is double-brooded with the first brood in late spring. The preferred habitat in the southeast US is suggested to be sandhills. In the Midwest, it is reported to inhabit savannas and pine barrens. It will readily visit many kinds of flowers, including the blooms of the host plant. Males are reported to visit moist soil.

Mottled Duskywing are about one-half the size of the other LA duskywings. Those individuals I have found were perched on either a stalk of tall grass or on the end of a tall dead weed. When disturbed, they all flew rapidly in a circular fashion, between four and eight feet above the ground, around their chosen perch (so fast as to be hard to follow). Eventually, all three returned to their original perch (or extremely close thereby). One was flushed several times but repeatedly returned to its perch. This behavior is unlike that of other duskywings found in Kisatchie NF. Sleepy and Wild Indigo Duskywings patrol more than perch. Horace's and Juvenal's Duskywings perch, but more so on the ground or on oaks. Also, they don't circle and return to a particular perch, but simply move off when disturbed.

My Records: I have found this skipper in two open fields within the NAT unit of Kisatchie. All three of my specimens were males taken in late Apr and were well worn, suggesting the first flight period is mid- to late Apr. I have walked those fields on other occasions later in the year without evidence of a later flight.

Distribution and Abundance: The Mottled Duskywing has been listed in 2015 as a Tier III, S3-ranked SGCN in LA. Beyond mine, other LA records are from central LA and the FL Parishes. The central LA record was from NAT (Kisatchie NF) in mid-Apr. It was first found by Strickland in the FL Parishes about four miles east of Greensburg in longleaf pine–scrub oak hills, where he found five species of *Erynnis* flying along with Dusky Roadside-Skippers and Dusted Skippers. He had numerous specimens from SHE within his collection, taken in late Mar to early Apr and mid- to late June. Other records from the eastern

Mottled Duskywing *(Erynnis martialis)* *(continued)*

portion of the state include the Global Wildlife Center and the Fluker area, both recorded in July. One Mottled Duskywing was reported on the 1997 Pearl River count (June). I anticipate it might turn up in north LA as it has been reported in several AR counties immediately north of CAD. It has also been reported from the region at Red Slough WMA in southeast OK.

Host Plant: New Jersey Tea, which has been suggested to be in decline across its former range for numerous reasons, including heavy browsing by deer. That particular plant is not common at Kisatchie NF, but I have found stands of it in the area of the two fields referenced above. The soil in the area is very sandy and seems to better support the food plant.

Zarucco Duskywing *(Erynnis zarucco)*

dorsal

ventral

dorsal

Description and Behavior: In most instances, the Zarucco Duskywing is extremely dark with no noticeable mottling either dorsally or ventrally. Also, the forewing is smaller and more pointed. In the fall, it is extremely hard to tell this skipper from a Horace's as both are very dark. One suggested differentiating

Zarucco Duskywing (*Erynnis zarucco*) (continued)

field mark is reported to be a light patch on Zaruccos, located about halfway along the leading edge of the dorsal forewing. I also look for reduced white/silver spotting on the forewing, both dorsally and ventrally, with the innermost spot missing, indicative of this skipper. Another factor to consider is whether there are any black locust seed pods (very large, black, and distinctive in their appearance) in the area, again indicative of this species.

Several sources report that in south FL (and possible elsewhere in the Deep South) this species can have light gray or even white fringes, making it look similar to the Funereal Duskywing. Strickland (1972) addressed a series of what he described as "possible *funeralis* X *zarucco* hybrids" from IBV, WFE, SHE, EBR, and CAM that "closely resembles *funeralis.*" The fringe was described as primarily brown, but the long scales were dirty white. One of these specimens is currently in the LSU Collection, taken in SHE in Aug, identified as "E. z. zarruco?" There were also two specimens within the LSU Collection with similar dirty/off-white fringe, collected and donated by Israel, from JEF (Sept) and WFE (Aug). Brou also had some brown-fringed specimens in his collection, taken during the same time frame. During personal conversations, Brou indicated that he was aware of Strickland's hypothesis but did not necessarily agree that such specimens represented hybrids based on such a small sample. I have included a picture of such a specimen seen at Northlake Nature Center in Aug.

My Records: This duskywing can often be found at mud or mud puddles. It is regularly present at Thistlethwaite WMA in the fall, most often at puddles along the dirt roads. I also have found it at Indian Bayou WMA at puddles on four-wheeler trails (July). The year 2013 seemed to be good for this duskywing as I saw specimens at four separate locations over the course of two weeks in mid-July.

Distribution and Abundance: Current records indicate a flight period from Mar to Nov, at times present in good numbers. There has been only one record from north LA. Southwest LA records are from Intracoastal City. Most records are from the NO region and the FL Parishes. It was common during Israel's Tunica Hills survey, but rare at his home in EFE. The flight times were May to Oct, although he had no records in July or Aug at his home. Other records include the Edgard area in the River Parishes region. Patton posted pictures on the LA listserv in mid-Aug of this species, found near the community of Pine in WAS

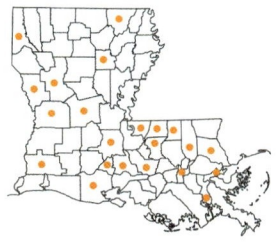

(this late record is not reflected in the distribution map). It has shown up on a few NABA counts, again primarily in the eastern portion of the state. On the Global Wildlife and Pearl River counts, it has been reported on five and seven counts, respectively, all in June or July. The highest numbers reported on the Pearl River count was fifteen in 1997, and then ten in 1994, attributed to the presence of plants in the Daubentonia family. Finally, it has been reported twice during the New Orleans count with eleven seen in June 2001.

Host Plants: It has been reported in LA feeding on black locust and hemp sesbania. Other members of the locust-tree family are also used, such as honey and water locust. In FL, it is reported to use herbs, vines, and shrubs in the bean family, including milk peas, indigos, bladderpod, Florida vetch, American wisteria, and occasionally wild indigo.

Funereal Duskywing *(Erynnis funeralis)*

ventral

dorsal

Description and Behavior: This particular dusky-wing is quite distinctive with a bright white dorsal and ventral fringe strip along the bottom of the hindwing. My experience suggests this duskywing is more attracted to flowers than the others (although all duskywings will stop at low-growing flowers). I have not found this one at mud and puddles like its cousins.

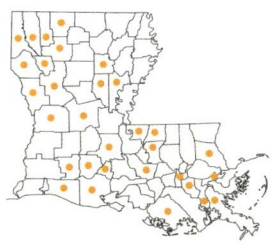

My Records: I have found it randomly in central LA and Acadiana, mostly in the fall, primarily Aug and Sept. My only eastern record was at Talisheek Pine Wetlands Preserve in early Aug. I have also seen it regularly in southwest LA in Oct and Nov. In Oct of 2012 and 2014, I saw several fresh individuals in the Pecan Island area of southern VRM. My only summer records have been Sicily Island Hills WMA in early June, Kisatchie NF in NAT in mid-June, Copenhagen Hills in late June, and Cooter's Bog in early July. My only spring record was early Apr at Kisatchie NF in NAT.

Distribution and Abundance: Strickland was the first to report this duskywing from WFE in early July. During his survey period, he found specimens in north LA (June), southwest LA (Aug and Sept), south LA (Grand Isle, Aug) and the FL Parishes (Aug through Nov). Other north LA records include CAD (occasional from early Mar, through the summer into late Oct) and RDR (Apr). During the Cajun Prairie survey, it was recorded from Aug through Oct, and there is a more recent, early Apr record from Clear Creek WMA. Records from the FL Parishes include Asphodel Plantation (June, July, and Aug) and Tunica Hills (June, Aug, and then Oct into Nov). In the River Parishes, it has been seen in Edgard. Israel (1981) considered it to be uncommon and felt there were two broods in the Tunica Hills region. In south LA, it has been reported sparingly (Feb). It was reported on two Grand Isle counts (May 2003 and Aug 2004), two Bonnet Carré counts (2008 and 2009), one New Orleans count (June 1997), one Pearl River count (June 1997), and one Cameron count (July 1992).

Host Plants: In southeast TX, it is reported to use rattlebush, vetch, indigo, alfalfa, and other legumes.

Wild Indigo Duskywing *(Erynnis baptisiae)*

ventral

dorsal

dorsal and ventral

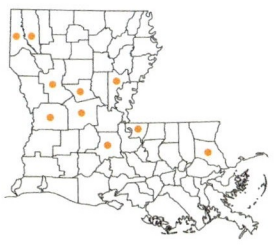

Description and Behavior: I find this skipper is most similar to the Sleepy Duskywing in appearance and habits; however, it is not as dark as a Sleepy, nor are its two rows of white spots on the lower ventral wings as bright. It is much more mottled than Zarucco Duskywings. This duskywing can usually be distinguished from both Horace's and Juvenal's by the lack of a white spot in the central forewing. Also unlike Horace's and Juvenal's, the males of this species patrol over areas of growing indigo, flying just above the grass tops. I have not seen the males actually perch on the indigo plants like Frosted Elfins will do; rather, they perch on tall stalks of grass or on patches of dirt. While Wild Indigo's flight is fast, it is more controlled, more specific than that of the two oak feeders.

My Records: I have found it primarily in central LA at the GRA, NAT, RAP, and VER units of Kisatchie NF, all in areas where yellow wild indigo was growing in profusion. Its first flight period is late Mar to early Apr, and then again during the last half of May into early June. I also found what might be a third brood flying in July to mid-Aug at locations in the Cajun Prairie region. Two good places to find it are in the large field next to the Catahoula Butterfly Garden or in the power-line cut that crosses Castor Plunge Road in the RAP unit. In early July, I found one at Cooter's Bog perched out in the middle of the bog. The only indigo plants I identified at that location were cream wild indigo plants, and the following June I witnessed a female oviposit on that species of wild indigo.

Distribution and Abundance: This particular duskywing has a more reduced presence in LA than its cousins, due to the limited distribution of its preferred host plant, wild indigo. It was listed in 2015 as a Tier II, S2/S3-ranked SGCN in LA. North LA records include Bodcau WMA in July and Eddie Jones Park, also in July. In the Acadiana region, there is an old record from the Jennings area. The LSU State Collection contains a specimen from WFE (Mar), and it has been reported on two Pearl River counts (1982, two seen, and 1984, three seen).

Host Plants: In the extreme western end of the FL Panhandle, caterpillars were found feeding on Florida hairy wild indigo and Apalachicola wild indigo. In southeast TX and the Texas hill country, it is reported to use wild indigos, lupines, crown vetch, rattlebush, and other legumes.

Common Checkered-Skipper *(Pyrgus communis)*

ventral

dorsal male

dorsal female

Description and Behavior: This skipper often is found flying with its cousin, the Tropical Checkered-Skipper. The best method of which I am aware to differentiate the two is to look closely at the fringed margin of both the forewings and the hindwings. If the fringe is primarily black with occasional small, white patches, it is the Tropical Checkered-Skipper. In contrast, if that fringe is primarily white with occasional small, black patches, then it is this skipper. The females are darker dorsally than the males, with the males showing much more white.

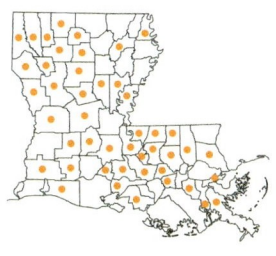

As spring turns to summer, this little skipper becomes more numerous. In Acadiana, during the dog days of July and early Aug, this is one of only a very few butterflies seen in numbers. By Sept, it is everywhere—disturbed areas, open fields, ditches, roadsides, wooded edges, even open woods. It is colonial, rarely seen as singles. It flies low, at grass level, often stopping to take nectar at low-growing flowers and mud. While moving from flower to flower, both males and females fly in a casual fashion, unusual for a skipper, and it can be easy to approach. Israel (1981) recorded it at eighteen different flower species.

My Records: This skipper is very common in the southwest LA and Acadiana regions. My records are from Apr to Nov, and it is clearly multi-brooded with possibly as many as five broods.

Distribution and Abundance: One of LA's most common skippers, it has records for every month of the year. In CAD, it is uncommon to common with records from mid-Jan to early Dec. In the central LA region, it has been found in Kisatchie NF at the NAT unit (June). Four broods have been reported in both Felicianas, from late Jan into early Dec, found in fields and waste and sandy areas. In south LA, both this species and the Tropical Checkered-skipper are present near Fort Jackson. It has shown up at least once on every NABA count in LA. On the 2000 Cameron count, a high of eighty-three was reported. On the Pearl River count, there were highs of ninety-five in 2008, eighty-two in 2007, and fifty-seven in 2006. The New Orleans count yielded seventy-one in 2005 and forty in 1993. The Barataria count had a high of fifty-three and forty-five in 2004 and 1999, respectively. Finally, the high on the Alexandria count was forty-nine in 2010.

Host Plants: It uses mallows, globe-mallow, and hibiscus as larval food plants.

Note: BAMONA has included the White Checkered-Skipper (*Pyrgus albescens*) as part of LA's fauna, probably based on records from Strickland, who reported this skipper as common throughout the state. That said, Strickland also noted there is a question about what species has been seen where. Using Tilden's criteria to differentiate between communis and albescens, Strickland (1972) states, "albescens-like specimens are found in the southwestern and south-central part of the State. As on [sic] moves north and east all sorts of interesting genitalic

Common Checkered-Skipper *(Pyrgus communis)* *(continued)*

variations occur, with communis-like specimens being taken in the more northerly reaches of the State." Burns (2000) has suggested these two species cannot be distinguished except by examining the male genitalia. Burns generated a map showing the distribution of both species in the US, which showed this skipper's range extending along the Gulf Coast into FL. Several NABA counts have reported "Common/White-Checkered Skippers," including Bonnet Carré (2005), Folsom (2003 and 2004), and Barataria (2009). I cannot tell them apart in the field and do not have the expertise to do genitalia evaluations. Therefore I have not attempted to distinguish which Common Checkered-Skipper records might have been this species.

Tropical Checkered-Skipper *(Pyrgus oileus)*

ventral

dorsal male

dorsal female

Tropical Checkered-Skipper *(Pyrgus oileus)*   *(continued)*

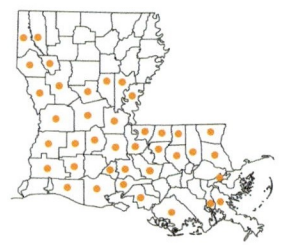

Description and Behavior: In the southwest portions of LA, particularly later in the year, this skipper becomes very common, often outnumbering its cousin. In behavior, the two species are indistinguishable.

My Records: My records are Mar into Nov (at Cypremort Point SP). I believe there are at least three broods (spring, mid- to late summer, and mid-fall), possibly a fourth in late fall. I have found it most often at Thistlethwaite WMA and Indian Bayou WMA. It is also regularly found in Kisatchie NF units in NAT, RAP, and GRA, but in fewer numbers. It is very common all over CAM in late fall.

Distribution and Abundance: The first specimens from LA were collected by Strickland (1972) in Baton Rouge and from southeast LA "along the coast." Later records encompass the entire state. It is uncommon in CAD, with records in Apr, July, and mid-Aug to mid-Nov. In the Acadiana region, it will fly from Feb into early Dec (Acadiana Nature Park). It is abundant at Asphodel Plantation from June through Oct. Other sightings include several NABA counts: Cameron (with a high of twenty-two in 1994); Pearl River (on numerous counts with a high of twenty-four in 1999); Global Wildlife (again, on numerous counts); New Orleans (several counts); Alexandria (on every count); Folsom, Grand Isle, and Barataria (several counts, highs of sixteen and thirteen).

Host Plants: In southeast TX, it is reported to use sida and other mallows as its larval food plant. In FL, it uses herbs in the hibiscus family such as broomweed, Indian hemp, and false mallow.

Common Sootywing *(Pholisora catullus)*

ventral

dorsal

SKIPPERS

Common Sootywing (*Pholisora catullus*) (*continued*)

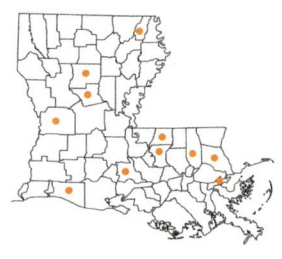

Description and Behavior: This small skipper could be confused with the Hayhurst's Scallopwing. If a close-enough look is allowed, the wings of this skipper are more rounded. Also, this skipper will have many more small white spots on the upper wings. Its flight is swift and low to the ground. The males appear to patrol, occasionally stopping at low-growing flowers. Generally, this skipper prefers drier habitat.

My Records: I've found it only at Indian Bayou WMA in July 2003 and June 2006, in an open area around the edge of a gravel road and parking area. This area is used by hunters, subjected to more traffic than most of the WMA, and is regularly mowed. I've not seen it since, despite repeated efforts to relocate it. Indian Bayou WMA would be the exception to this skipper's general preference for drier habitat, which may be a possible explanation why it is not regularly seen there.

Distribution and Abundance: Although it has shown up around the state, it is not a common skipper. In north LA, Brou had two specimens from ECA. In central LA, there are two specimens in the LSU State Collection from GRA and VER, both taken in Apr. In southwest LA, it has been found at Peveto Woods (Apr). In the FL Parishes, it has been recorded in EBR (Mar), EFE (Oct), and STA. It has also been reported on the Pearl River count in 1977 (twenty-eight seen), 1994 (fifteen seen), 1998, and 1999, and on the New Orleans count in June 1997. Phillip Wallace reported a sighting in early Aug 2017 and then posted pictures on the LA listserv in mid-Aug of this species, found near the community of Pine in WAS (this late record is not reflected in the distribution map). These spotty LA records match its distribution in MS, AR, and east TX.

Host Plants: In the Texas hill country, it is reported to use pigweed and other members of the goosefoot family. On the East Coast, it is reported to use cockscomb.

Swarthy Skipper *(Nastra lherminier)*

ventral

dorsal

SKIPPERS

Swarthy Skipper *(Nastra lherminier)* *(continued)*

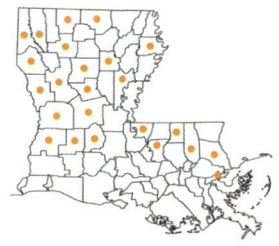

Description and Behavior: The Swarthy Skipper is a small, nondescript skipper that frequents LA's pine forest. Klots (1951) described it as neutral, dull, drab, and undistinguished. One visual characteristic to look for is pale yellow tracing of the veins on the lower ventral wings in fresh individuals. If those tracings are present but with essentially no other field marks, then more probably than not it is this skipper. None of the Swarthy I have seen in LA possessed any of the spots some sources suggest can be present. I also noted that none of the Swarthy specimens in Strickland's collection had any spots.

My Records: I first found Swarthy Skippers in May, during the 2008 Kisatchie NAT unit count. Several small brown skippers with absolutely no discernible identifying marks were flying in the grass along an old dirt road through an area of open pine woods and were eventually identified as Swarthy Skippers. Since then, I've also found them in identical habitat: Sicily Island Hills WMA (June–July), Kisatchie NF in RAP and VER (Mar–Sept), Cooter's Bog (May), and Catahoula Butterfly Garden (Apr, July, and Oct).

Distribution and Abundance: There are limited records for this small, low-flying skipper, although it is probably more widespread than those records reflect. There are four sightings (May, June, July, and Nov) from CAD. Other north LA locations include Kisatchie NF in WIN (Sept–Oct), Coney (May), Bodcau WMA (July–Sept), and Roy Quad (June). In the FL Parishes, it has been documented in the Tunica Hills region in Apr, June to Aug, and Oct, where it is uncommon over what appear to be two broods. Other records from that region include EBR (May) and SHE (July). It has shown up on several NABA counts to include Catahoula (Sept with fifteen seen), Pearl River (June), New Orleans (June), and Global Wildlife (July).

Host Plant: In southeast TX, it is reported to use little bluestem grass.

Neamathla Skipper *(Nastra neamathla)*

ventral

dorsal

SKIPPERS

Neamathla Skipper *(Nastra neamathla)* *(continued)*

Description and Behavior: The Neamathla Skipper was recognized as a species in 1923, but its range boundaries are still imprecisely known. This skipper and the Swarthy Skipper exhibit similar appearance and behavior. They are the quintessential grass skippers, typically found in grass along the sides of roads or trails. They fly low, down in the grass. On occasion I have found them sunning on dirt roads with closely mowed grass. They will take nectar at flowers, and seem to like low-growing asters in the fall. The one regularly reported distinguishing characteristic between Swarthy and Neamathla skippers is that the veins on the ventral hindwing are distinctively lighter colored on the Swarthy Skipper. Some sources mention that the spots on the upper forewing of Neamathlas are a trifle more prominent than on Swarthys. While none of the Neamathla Skippers I inspected from LA (to include those collected by Strickland) possessed any spots, I found several Neamathla Skippers in Jackson County, MS, that possessed the dorsal spots, and the specimens pictured by Linda Cooper from FL also show those spots. I would suggest habitat is another method by which to distinguish this skipper from the Swarthy Skipper. While both seem to prefer open pine woods, in all of the instances where I have found this species, the specimens were flying in low-lying grassy areas with standing water. In contrast, typical Swarthy habitat is more upland, dry grassy areas. I recognize some of the records by others (noted below) may not fit this distinction, and would suggest more research on the range of this skipper is required.

My Records: While in CAM one May, in a damp field along Highway 27E, I caught a small brown skipper flying just above the grass. Tentatively identifying it as a Swarthy, I wondered if it might instead be a Neamathla Skipper. In Oct that same year, I returned and found seven more specimens at the same location. These specimens matched the Neamathlas in the Strickland collection from CAM. It can be common across CAM in Oct. I also found it at the Freshwater Bayou Lock Rec Area in late Oct. In the east, I have seen good numbers at times during Apr, Aug, Sept, and Oct at Abita Creek Flatwoods Preserve and Lake Ramsey Savannah WMA, as well as in a wet power-line right-of-way near Enon in WAS. In Mar 2017, I found several specimens at the Persimmon Gully Preserve in CALC and CC Road Preserve in ALL which I felt were this species. The habitat matched the generally wet, open grassy habitat where I have found this species in CAM and STA. Because none of the specimens at

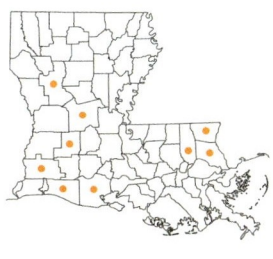

either location had any dorsal spots, I decided to send specimens to Roever for genital examination (along with Swarthy specimens from Dove Field in VER). Roever confirmed my diagnosis, noting that even though the ALL specimens had no dorsal spots, they were Neamathla Skippers. It is clearly multi-brooded, probably having four broods (spring, early summer, late summer, and fall), possibly more.

Distribution and Abundance: The first LA record was a male in early Sept 1967, in CAM. Other specimens were subsequently recorded in CAM in late Apr, and in mid-Sept. Species identification was made by genital examination. Neamathlas have been reported during several CAM counts, centered at Peveto Woods, including in July 1992, when Strickland and Ross reported two, indicating identity was verified by capture. In his 1972 presentation, Strickland reported that no Swarthy specimens had been collected along coastal LA, and no Neamathlas had been collected inland. Since then, eleven Neamathlas were reported during the 2000 Alexandria count (Aug), a count held far from the coast. Because the count coordinator, Marty Floyd, wasn't sure about the identity, he provided specimen(s) to Strickland who, through dissection, identified the specimen(s) to be Neamathla based on the male genitalia. During personal conversations with Strickland in 2010, he advised that he has also caught a Neamathla in RAP. Brou had specimens from NAT. This skipper has been recorded several times in the FL Parishes near Abita Springs. Also, one was seen in the July 2000 Global Wildlife count (as well as two Swarthy Skippers).

Host Plants: In southeast TX, it is reported to use Johnson grass and dallisgrass.

Clouded Skipper *(Lerema accius)*

ventral

dorsal

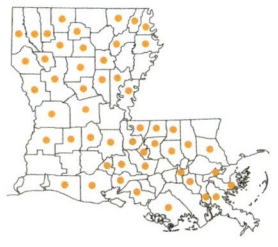

Description and Behavior: This skipper typically flies low, at grass-top level. It is a regular visitor at numerous kinds of flowers, such as Brazilian vervain and ironweed. Israel (1981) found it at fifteen different kinds of flowers during his thesis survey. It is a fast flier, even when not disturbed. The males race around, chasing each other before returning to perch. The females are typically larger, have spots on the upper wings, and are also easier to approach while taking nectar.

My Records: My sightings are from Apr to Nov. It can be seen regularly at Thistlethwaite WMA throughout its flight period from Apr into Sept, occasionally surviving even into Nov. It is also common at the Kisatchie units in RAP and GRA. A good place to look in the latter unit is along the main road in Stuart Lake Recreation Complex. I've found it to be very common during Oct all over CAM. It is clearly multi-brooded, with probably three, possibly four, broods. I added a late record in mid-Aug from the Enon/WAS area that is not reflected in the distribution map.

Distribution and Abundance: This is a common skipper in LA, flying much of the year, although not seen as often early in the year. The earliest record of which I am aware was in SBN on Jan 23. It was also reported in late Feb in LAFA at Acadiana Nature Park. In CAD, there are records from early Mar to early Dec, common to abundant from late July into mid-Nov. It can be abundant at Asphodel Plantation, flying from Mar until Dec, most common in Sept and Oct. It is also common in WFE in open fields, from late Mar to early Dec. River Parishes records include sightings near Edgard. It has been reported on many NABA counts, including Shreveport (with a high of forty-four seen in Sept 2010), Catahoula (with ten seen in 2010), Alexandria and Pearl River (multiple times), New Orleans (multiple times), and Bonnet Carré and Global Wildlife (both twice).

Host Plants: Two grasses, Indian corn and Johnson grass, have been specifically listed as used in LA. In southeast TX, it is reported to use rustyseed paspalum, St. Augustine, and other grasses. In FL, it uses different grasses including redtop, purpletop, gama grass, Johnson grass, Guinea grass, barnyard grass, elephant grass, and sugarcane.

SKIPPERS

Least Skipper *(Ancyloxypha numitor)*

ventral

dorsal

Description and Behavior: Slow and feeble in flight, it is most often seen at or near standing or slow-moving water. It appears to be colonial, more often found in numbers rather than as singles. Its slow flight is an excellent way to distinguish it from the swift-flying Southern Skipperling. Despite its slow flight, it can be quite evasive, given its habit of actually flying down in the grass and/or brush.

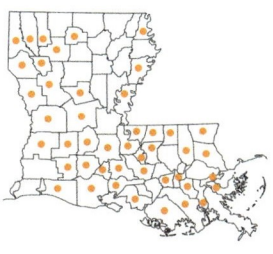

My Records: My records are from Apr to Nov. It is a regular at Thistlethwaite WMA. I can count on seeing it each year during midsummer in the middle of the city of Lafayette, feeding on low-growing yellow lantana. I have also found it all over CAM and southern VRM in Oct into early Nov while the weather remains warm. My latest record is Nov 12, with twenty-seven seen at the Rip Van Winkle Rookery in IBE. Along the Mississippi River corridor, I found it at the boat launch at Bayou Cocodrie NWR in CON. It is multi-brooded, and seems to fly continuously, at Thistlethwaite WMA.

Distribution and Abundance: It is common in CAD, from May through Oct. Other north LA locations include the Bayou Pierre Unit (Nov), Bayou Macon WMA (Aug), Bodcau WMA (Aug), Caney Lake Rec Area (late July), and Newlight (Aug). Records from Kisatchie NF in NAT and RAP (Aug–Sept) confirm its presence in central LA, and Acadiana records include the Cajun Prairie region (Apr–May). In southwest LA, it has been found at Lacassine NWR (Nov and Dec). In the east, there are records at Grand Bay, Sherburne WMA (Nov), near Erwinville and Greenwood Park (throughout the summer into Nov). It is in both of the Felicianas, flying from Apr to Nov, described as locally common in grassy, moist areas. Brou had multiple specimens, including at Edgard. In southernmost LA, it has been recorded twice (June and Aug) in the Galliano area. It can be abundant in the right habitat. For example, it has been reported on multiple Barataria counts with highs of 117, 88, and 25, all in June. It is also regularly reported on the New Orleans count (again, in June), once with a high of 13. Other counts on which it has appeared include Pearl River, Cameron, and Bonnet Carré.

Host Plants: In southeast TX, it is reported to use marsh millet, bluegrass, and rice. Its use of rice has been reported in VRM.

Southern Skipperling *(Copaeodes minima)*

ventral

dorsal

Description and Behavior: With its habit of flying low to the ground, LA's smallest skipper can be easily missed. It has a swift flight, but it rarely flies very far. Males perch on stalks of grass, and are easily observed once discovered. It might be confused with the Least Skipper, but the two are easily separable. First, the

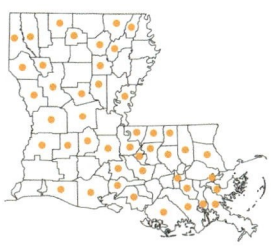

flight pattern of this skipperling is very different, so swift that at times it is difficult to follow. Second, the Southern Skipperling is smaller and shows less black scaling than the Least. Like the Least Skipper, it is colonial, so where there is one, there should be more. It is multi-brooded, with at least three broods (spring, summer, and fall), possibly even a fourth in late fall.

My Records: I find it regularly throughout Thistlethwaite WMA (even in mid-Feb). Once, in late Aug, I witnessed fifty-plus flying in Sherburne WMA, feeding on frog fruit around the boat dock at Ramah in IBV. It prefers open, moist areas such as in and around ditches. I also found it at both Duralde Prairie and Eunice Cajun Prairie. It is a regular at the GRA unit of Kisatchie NF, both on the main road at Stuart Lake Recreation Complex and in the Catahoula Butterfly Garden. I have found it to be common in CAM in Oct and Nov, wherever there are grassy areas near ditches and/or canals, including Peveto Woods, Cameron Prairie NWR, and Sabine NWR. Other southern locations include Freshwater Bayou Lock Rec Area (late Oct) and Cypremort Point SP (mid-Nov).

Distribution and Abundance: Lambremont (1954) called it the Tiny Skipper, most often found in open, sunny fields. In CAD, it is common from mid-May to mid-Nov. It has also been found at Bayou Pierre Unit (Oct), Cane's Landing, and Bodcau WMA (both in Aug–Sept). In central LA, it has been recorded at the Kisatchie units in VER, RAP, and NAT (Aug to Oct), and, in southwest LA, at Intracoastal City. There are multiple records mid-state, along the Manchac Road and near Erwinville and Greenwood Park, flying throughout the summer and into Nov and Dec. In EFE, it is rare in Apr to May and in July, and in WFE it is uncommon in grassy areas, with records in May and from July to Nov. FL Parishes records include Greensburg, and River Parishes records include Edgard. In southernmost LA, it has been reported at Grand Isle and Galliano, flying from May to June and Sept to Nov. It has been regularly reported on the Bonnet Carré count (typically Aug), with a high of fifty-four in 2000. It has also been regularly reported on the New Orleans count (June), with a high of thirty-three in 2002. Other counts include Barataria (June), Pearl River (June), Cameron (July), and Global Wildlife (Aug).

Host Plants: In both LA and southeast TX, it is reported to use Bermuda grass and perhaps other grasses.

SKIPPERS

Fiery Skipper (*Hylephila phyleus*)

ventral male

dorsal male

dorsal female

ventral female

Description and Behavior: This is probably the most common grass skipper in LA, often found in disturbed open areas, city gardens, and even sports complexes. The males perch on stalks of grass and often chase each other with a swift flight. Both sexes visit flowers like clover and frog fruit but remain wary and hard to approach. The females are larger

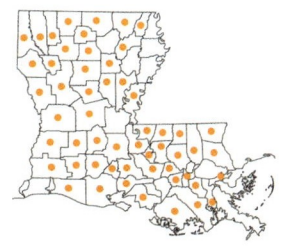

and a duller gray, while the males are a brighter tan. This skipper might be confused with the Whirlabout and the male Sachem. The Whirlabout is smaller, and the spots on its ventral hindwings are larger, fewer, and darker brown. The male Fiery Skipper and the male Sachem are very similar and fly together in western LA. The best distinguishing feature is the large black bar on the dorsal forewing of the Sachem male, and ventrally, the Sachem male's hindwing has lighter patches than the darker, small spots of the Fiery. Occasionally, males will have very few, if any, spots and can be confused with the Delaware Skipper, but dorsally, they are very different.

My Records: I have records from late Feb and then Apr to Dec. There are four broods: spring, midsummer, early fall, and then late fall. In CAM, it flies continuously throughout the summer, very common to abundant from Sept to early Nov. It is the most common fall butterfly at the Catahoula Butterfly Garden, with large numbers feeding on the yellow and purple lantana with the males engaging in a never-ending game of chase. It is also common in and around Thistlethwaite WMA in the fall.

Distribution and Abundance: There are records from every section of the state. It is common to abundant in CAD from early Apr into Nov with a single sighting in Dec. During the 2009 Shreveport count, more than one hundred were counted at a small cattle ranch. During the Cajun Prairie survey, this skipper was seen in four parishes. Recorded as common, sometimes abundant, in both of the Felicianas, it flies from Jan to Dec. Brou (1974) found seven in his ultraviolet light traps in SJB. In southernmost LA, Grand Isle (one record) and Galliano records are from Feb to Mar, May to Aug, and Oct to Nov.

Host Plants: It uses various grasses, including crabgrass. In southeast TX, it is reported to use lawn grasses. In FL, it uses Bermuda and St. Augustine.

Cobweb Skipper *(Hesperia metea)*

dorsal male

ventral male

ventral female

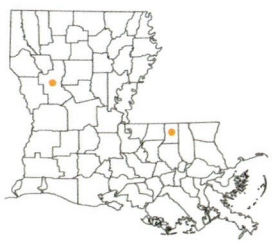

Description and Behavior: This is a small skipper that is restricted in host plants, habitat, and flight period. It flies only in early spring—late Mar into early Apr. Cech and Tudor (2005) describe its habitat as dry or successional pine barrens, pine-oak savannas, open scrub, and burn sites. Per Stichter (2015), in the eastern portion of its range, it is a single-brooded, dry upland skipper, which prefers sandy or rocky sites, "most likely to be found in large established stands of little bluestem rather than . . . in thin stands stretching along roadsides." It is reported to be fire-adapted.

It has been occasionally reported in MS, taken in late Mar to late Apr, generally rare and scattered. It is also known from southwest AR in Miller County and in the piney woods region of northeast TX, so it could also be found in similar areas across north LA. Killian Roever verbally reported that this skipper occurs in the same areas as Dusted Skippers (dry, open-pine woodlands where its food plant grows), but about seven to ten days earlier. He described them as much more wary than Dusted Skippers (which are wary themselves). He also advised that it was attracted to blackberry blooms and white clover. It is colonial. The males stake out territories on dirt roads. Its flight is reported to be weak, with the males perching on or near the ground. It was listed in 2015 as a Tier III, SU-ranked SGCN in LA.

My Records: I have had no experience with this skipper.

Distribution and Abundance: Strickland (1972) was the first to find it with a dozen recorded between March 24 and April 8, a few miles east of Greensburg in habitat identified as longleaf pine-scrub hills. (The area is no longer open pine woods.) Strickland described it as extremely dark with the males showing only "a trace of yellow adjacent to the stigma." The females were reported to be "featureless except for small white spots on the forewing beneath." Strickland donated one male and one female from this site to the LSU State Collection. Roever verbally reported finding this skipper in mid-Mar the following year, in Kisatchie NF in NAT along the Scenic Byway (before the road was straightened and paved).

Host Plants: On the East Coast, it is reported to use little bluestem and big bluestem. Roever reported to me that it used broomsedge.

Meske's Skipper *(Hesperia meskei)*

ventral

dorsal male

dorsal female

Description and Behavior: This skipper's range is primarily across the southern US with isolated colonies in central AR and east TX. The subspecies in the piney woods region of east TX is *H. m. meskei*. The habitat in Kisatchie NF is identical to that of east TX, which is less than fifty miles away, and I believe the population in Kisatchie is an eastward extension of that subspecies. I've found no record of this skipper in MS, so it does not appear the LA population came

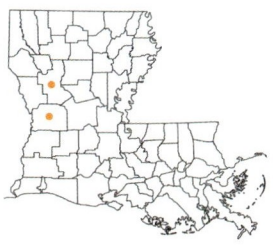

from the east. A rare skipper in LA, it was listed in 2015 as a Tier III, S1-ranked SGCN. The meskei subspecies is reported to have two broods, May to June and Sept to Oct. It is found mainly in dry, sandhill pinelands and is well adapted to fire-prone habitats. Both sexes readily go to yellow, blue, and white flowers. I have found it favors a tickseed member of the Coreopsis family, but will also take nectar at yellow flowers in the coneflower and sunflower families, all of which grow in roadside ditches and back in the pines. It is a fast flier with an audible "buzz."

My Records: My records are all from the NAT unit of Kisatchie NF, primarily along a dirt road through dry, open pine forest with several species of tall grasses growing beneath the pines. In mid-June 2009, I saw 34, a surprising number, or so I thought at the time. During the 2010 Kisatchie NF count, also in mid-June, a record was set for the most Meske's Skippers tallied during a NABA count, 42. In June 2013, I found specimens in other areas of the unit, including moister habitat with significantly more deciduous trees. In mid-Oct 2014, I found both sexes in the area of a pitcher-plant bog within the NAT unit. Within the next week, more were found in the same area, confirming the existence of a second brood.

Distribution and Abundance: Charles Bordelon first found this skipper in LA in Kisatchie's NAT unit in the early 1990s. Kreg Ellzey (Tuttle 2003) reported a total of 18 individuals in that same unit, between June 15 and 23, 2003, at mountain mints and ironweed. On June 20, 2013, Trahan reported 140 in that unit along the Longleaf Vista Road at patches of mountain mint and a yellow tickseed species. The only confirmed sightings outside of NAT has been in mid-June 2015 and 2016, at Dove Field Rec Area in the VER unit of Kisatchie NF. Combined records indicate the June flight extends from about June 10 to July 1, peaking around June 20. The Lepidopterists' Society's Season Summary for 1996 had a report of *meskei*, caught in Sept of that year by Cunningham in northern TER, but Cunningham verbally advised that the skipper in question was, in fact, a male Yehl Skipper. I suspect this is the southern LA record referenced by Glassberg (1999).

Host Plants: Its larval host plants have been identified as little bluestem and arrowfeather threeawn. Both can be found in central LA along the edges of pine forests and disturbed areas, particularly west of the Mississippi River.

SKIPPERS

Sachem (*Atalopedes campestris*)

ventral female

dorsal female

dorsal male

ventral male

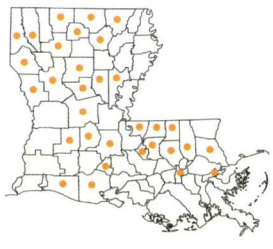

Description and Behavior: This midsized skipper frequents open fields, where the males perch and chase, dashing after intruders into their area before returning to their perch. Males perch on stalks of grass or on bare spots. Both sexes readily visit flowers and seem to particularly favor frog fruit. Although generally larger, the males can be confused with the more common Fiery Skipper, particularly as their behavior is similar. The large black bars on the males' dorsal forewings are diagnostic, but those bars are not typically visible when the skipper rests with its wings closed over its back. Females resemble Leonard's Skippers, which have been recorded in LA only once.

My Records: I have not found this skipper to be common. I have seen it in CAD in July, and at the Catahoula Butterfly Garden in Oct. I found one at a lake north of Thistlethwaite WMA in Oct, and then another actually within Thistlethwaite WMA in May. I saw two at Sicily Island Hills WMA in June. To the south, I've also seen it several times in CAM, including Peveto Woods and at Grand Chenier near Rockefeller NWR in the fall. I also saw several east of Pecan Island in late Oct.

Distribution and Abundance: Combined records include most of the state. In CAD, there is one Apr sighting; then it becomes uncommon to common from late May to mid-Nov. Other north LA records include Bodcau WMA (May–Apr and July–Sept), the Pintail tract of Red River NWR (Oct), Sparta Quad (July), and Kisatchie NF in WIN (Sept). In the central LA and Acadiana regions, it has been reported at Vernon Lake (Oct) and in the St. Landry area. In mid-state, it has been found near Erwinville (Aug) and Greenwood Park (Aug to Oct). It was reported to be rare in both of the Felicianas during Israel's 1981 survey, with flights from Mar to Apr and then June to Oct, over three broods. Other sightings in the FL Parishes include Baptist (Aug), near Greensburg (Mar), and the Global Wildlife Center. River Parishes records include Edgard. It has shown up on five NABA counts, at Alexandria (Oct 2011, Sept 2012, Sept 2013, and Oct 2016), Catahoula (Oct 2013), Folsom (Oct 2003 and 2004), Fort Polk (Sept 2013), and Cameron (Aug 2002 and 2003).

Host Plant: There is an old report from June, in LAFA, of this skipper feeding on Bermuda grass. In fact, Mather and Mather (1958) suggested it could be a pest where this grass is cultivated. In southeast TX, it is reported to use Bermuda grass as well as others.

Tawny-edged Skipper (*Polites themistocles*)

ventral male

dorsal male

dorsal female

Description and Behavior: Lambremont (1954) be-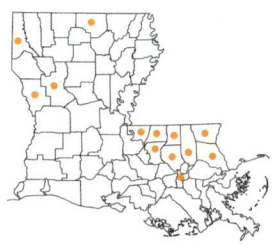lieved this skipper was "well distributed throughout" LA, although updated records are from only twelve parishes and would suggest otherwise. It is a smaller skipper, similar in size to the two broken-dashes. It tends to fly low to the ground, moving in and out of the taller grass. The males perch on grass stalks and, when disturbed, do not move far before again landing. I have also found males (primarily) perched on tall yellow Coreopsis flowers, and females taking nectar at blooming liatris and a low-growing blue flower I suspected was in the Ageratum family.

My Records: I did not see this skipper in LA until 2013, when I found it in good numbers at Abita Creek Flatwoods Preserve. I have since seen it at Talisheek Pine Wetlands Preserve, Lake Ramsey Savannah WMA, and in a power-line right-of-way near Enon in WAS. The habitat at all four is wet, pine Flatwoods savannah. There appear to be at least three broods, Apr–May, July–Aug, and Sept–Oct.

Distribution and Abundance: In CAD, there are records from late May into early July and again in mid-Sept. In central LA, it has been recorded in Kisatchie NF in NAT in Apr and May. In the east, records include Waddill Wildlife Refuge (July and Oct), near Greensburg (May to July), and Honey Island Swamp WMA (Sept). It has been sighted in SJB in Edgard. It has been reported as rare in EFE and WFE, with sightings in May and June in the former and July through Oct in the latter. In LIV, it was present near Holden (Apr). It has been recorded on the NABA count at Global Wildlife in July 1997–2000, but not since. It was also reported once (June 2005) during the Shreveport count, and once during the Pearl River count in 2009.

Host Plants: On the East Coast, it is reported to use grasses and occasionally sedges, including panicgrasses, bluegrasses, and crabgrasses.

SKIPPERS

Crossline Skipper *(Polites origenes)*

ventral male

dorsal male

ventral female

dorsal female

Description and Behavior: This skipper flies low to the ground in disturbed areas such as roadside ditches and open fields. Both its habits and appearance are similar to the Tawny-edged Skipper. The females can be particularly difficult to distinguish. The Crossline often shows a vertical column of pale spots on the ventral hindwing, although the spots are not always present.

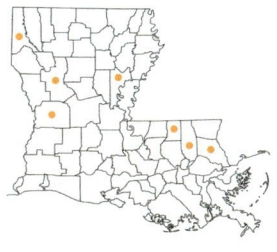

My Records: I've seen this skipper in two locations within the NAT Kisatchie unit in late Apr and early May and in the VER unit in Oct. I found a female at Abita Creek Flatwoods Preserve in mid-Aug 2013 that was noticeably different from the other Tawny-edged Skippers seen that day and which more closely resembled the Crossline females seen in NAT, but because I saw no other Crosslines, I remain hesitant to call it one.

Distribution and Abundance: Lambremont (1954) referred to it as *P. manata-aqua* and reported it as rare in the FL Parishes, found in moist, grassy openings in pine flats, all in Sept. Strickland had field records for this skipper and the Tawny-edged, near Greensburg (May and July). He also recorded specimens of both species in CAT, but his notes reflect that he was not sure which were which. None of these records were mentioned in his unpublished manuscript, leaving me to wonder if he was certain of his identifications. In the intervening years, I've seen it recorded only in north LA at Eddie Jones Park in May and on the 2016 Alexandria count (Oct). This skipper has been reported common and generally distributed in MS, from Apr to Oct. It has also been recorded in southwest AR and northeast TX, along the LA border. As it has been found regularly both east and west of LA, I expect it to be more present than the above records would reflect.

Host Plants: On the East Coast, it has been reported to use purpletop tridens, little bluestem, and mannagrass, possibly sedges.

Whirlabout *(Polites vibex)*

ventral

dorsal female

dorsal male

Description and Behavior: Occasionally abundant, the Whirlabout is typically less common across LA than the similar Fiery Skipper. It is most common in areas of open pine flats. This skipper is a fast flier, generally flying low, just above the top of the grass; males perch on grass stalks, buzzing after anything that flies past. A wary skipper, it readily visits many types of flowers (particularly lantana and Brazilian vervain), where it is easy to approach. The sexes are

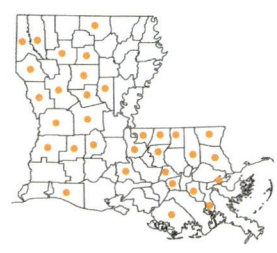

dimorphic, with the females being duller ventrally, showing almost a greenish tint as opposed to the bright tan males. Dorsally, the female is brown, appearing somewhat like the Little Glassywing.

My Records: I have seen this skipper most often in the central LA region, where I believe there are three broods, Apr–early May, July–Aug, and Sept–Oct. Despite once finding a faded female in the NAT unit of Kisatchie NF in mid-June, I'm not sure if there is a June brood. It can be common in and around the Catahoula Butterfly Garden, particularly in late fall. Sightings at Cooter's Bog were in May, July, and Aug. I have seen it occasionally at Indian Bayou WMA in Aug. I have also found it in CAM in late fall (as late as Nov), in areas where Bermuda grass is growing. In the FL Parishes, it is common from Aug to Oct at Abita Creek Flatwoods Preserve and Lake Ramsey Savannah WMA. It can be somewhat common in the spring in the Weyanoke area, in open areas within the forest where grass grows.

Distribution and Abundance: In CAD, records reflect a single sighting in Apr and then occasional sightings in June–July and Sept–Oct. Other north LA records are in Kisatchie NF in WIN, Sparta Quad (July), and Bodcau WMA (Aug–Sept). In central LA, it has been recorded in Kisatchie NF in RAP (Aug–Sept) and during the Catahoula count (Oct) with fifty reported. Eastern records include EBR from July through Oct, Grand Bay, and Edgard. It flies from Apr to Nov, over three broods, in the Felicianas, locally common in open fields. It has also been seen on numerous southern NABA counts: Pearl River, multiple years (all June) with a high of twelve; Barataria, two years (June); New Orleans, multiple years (all June) with highs of twenty-three and nineteen; Global Wildlife, multiple years (July and Aug) with a high of ten; Bonnet Carré, two years (Aug); Grand Isle, twice (Mar and Apr); Cameron, one year (July); and Folsom, three years (Sept and Oct) with highs of sixty and thirty.

Host Plants: It feeds on grasses. In southeast TX, it is reported to use grasses such as St. Augustine, Bermuda, and dallisgrass. In FL, it also uses crabgrass.

Southern Broken-Dash *(Wallengrenia otho)*

ventral

dorsal

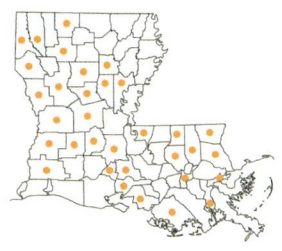

Description and Behavior: The Southern Broken-Dash is smallish, similar in size to the Whirlabout. It is lighter colored than its cousin, the Northern Broken-Dash, more tan than brown. Both possess the reverse "3" design on the ventral hindwing, but it typically is easier to see on this one than on its cousin. It has multiple light-yellow patches on both the dorsal and ventral forewings, which are larger and more yellow than those found in the Northern. When it has become worn, with the reverse "3" smudged or missing, it can be difficult to identify. For more identification keys, see the Little Glassywing account. It can be seen throughout the state from early spring into the fall over multiple broods, probably three (late spring, summer, and fall), possibly a fourth in late fall (Oct).

My Records: I have found it to be common to abundant at several locations: Sicily Island Hills WMA (early June), Copenhagen Hills (mid-June), Cooter's Bog (May), Bayou Teche NWR (early July), Tickfaw SP (Oct), Thistlethwaite WMA, and the NAT unit of Kisatchie NF (the latter two from Apr to early Sept). I have also found it in the Kisatchie units in RAP, VER, and GRA, as well as in a power-line right-of-way near Enon in WAS.

Distribution and Abundance: It is uncommon to common in CAD from early May to early July and then late Aug into early Oct. Locations include Walter Jacobs Nature Park, Kisatchie NF in WIN (Sept and Oct), Roy Quad (June), Corney Lake Unit (late May), and Bodcau WMA (May–Sept). In central LA and Acadiana, it has been found at Kisatchie NF in RAP (Sept), Chicot SP (Sept), and Acadiana Nature Park (Apr). It flies in WFE from early Apr to mid-Oct, uncommon, around flowers in old fields. In the River Parishes, it was recorded in the Edgard area. It has shown up during several NABA counts including New Orleans (eight counts, all in June), Barataria (two counts in June), Shreveport (two counts, June and Sept), Pearl River (one count in Aug), Catahoula (one count in Sept), and Alexandria (one count in Sept).

Host Plants: It has been reported to feed on grasses such as crabgrass and St. Augustine.

Northern Broken-Dash *(Wallengrenia egeremet)*

ventral

dorsal

ventral

Description and Behavior: This skipper was not recognized as a full species until the mid-1980s. It is less common than the Southern Broken-Dash, and a contributing factor toward the limited number of LA sightings for this skipper might be confusion between it and its cousin. This skipper is much darker brown, while the Southern Broken-Dash shows more tan color with more dis-

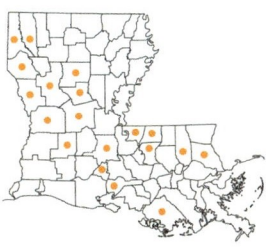

tinctive spotting. Those spots are whiter in color on the Northern species. For more identification keys, see the Little Glassywing account.

My Records: I have seen it in SLA, LAFA, NAT, RAP, GRA, CALD, STA, and WFE during late Mar to June and again in Sept. The latest I have seen it was October 19. It is most reliably seen at Thistlethwaite WMA; however, it is never as common as its cousin.

Distribution and Abundance: The first LA report of this species was for the 1992 Pearl River count (June). Other single sightings were reported on that count in June 1994, 2000, and 2002. It is uncommon in CAD with sightings from early May to early July and early Sept to early Oct. Other north LA locations include Bodcau WMA (Sept), Cane's Landing (Oct), and within Kisatchie NF in WIN (same months). At Asphodel Plantation it is common, flying from Apr to Oct. In southernmost LA, it has been recorded in TER. Patton posted pictures on the LA listserv in early Aug of this species, found near the community of Pine in WAS (this late record is not reflected in the distribution map). It was reported on four other NABA counts: Global Wildlife (July 1999), Thistlethwaite (June 2008), Alexandria (Sept 2010, Oct 2011, and Sept 2012), and Shreveport (Sept 2010).

Host Plants: On the East Coast, it is reported as using large panicgrasses such as switchgrass and deertongue.

SKIPPERS

Little Glassywing *(Pompeius verna)*

ventral

dorsal male

ventral female

Description and Behavior: The Little Glassywing is not habitat specific, flying in the hardwood regions of SMA, SLA, and LAFA and the pine woods of central and east LA. It prefers open, grassy areas, roadside ditches, pipeline cuts, and forest edges. With a swift flight, it is best viewed at flowers, where it will nectar quite readily at a number of low-growing wildflowers. When fresh, Little Glassywings have a ventral purple sheen that is distinctive, but when worn, they can be hard to recognize. Also, the glassine window on the forewings (when large) is distinctive. Unfortunately, the purple sheen wears quickly, and the glassine

ventral

window can be rather small on some male individuals. This skipper can be confused with Northern Broken-Dashes, Duns, and Clouded Skippers. Northern Broken-Dashes can be very dark in LA and have whitish markings in the center of the ventral forewing, but those spots are typically smaller than those on the Little Glassywing and are not glassine. Male Duns have a gold-colored head. Female Duns have fewer and smaller white spots on the dorsal forewing and do not have distinctive whitish spots in the center of the ventral hindwing. The Clouded Skipper is typically larger (particularly the females) and has white "clouding" along the outer edge of the ventral hindwing. The Clouded Skipper male can be smallish in the spring, and when so worn that the ventral hindwing clouding is lost, it can look like a worn Little Glassywing, but the wings on the Clouded males are narrower and longer than the Little Glassywing's short, relatively stubby wings.

Little Glassywing *(Pompeius verna)* *(continued)*

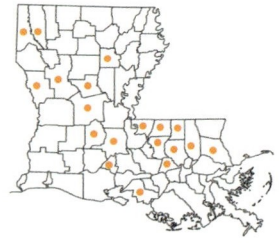

My Records: I have found it in the Acadiana region from Apr until Aug, in the central LA region from Apr until Oct, and in the Felicianas in Apr and then again in Aug and Sept. Specific locations include the RAP unit of Kisatchie NF (along low areas of Castor Plunge Road), Copenhagen Hills, Catahoula Butterfly Garden, and Tickfaw SP. It has been recorded several times during the Thistlethwaite WMA count (June 2007, June 2008, and May 2010). I believe there are two broods, in the spring and fall, with a possible partial brood in midsummer.

Distribution and Abundance: The Little Glassywing was first reported by Strickland (1972) in 1967. He described its typical habitat as "along roadsides through mixed deciduous woods." It has been found in CAD (Eddie Jones Park) occasionally, with records from May through July and in Sept and early Oct. In central LA, it has been found in the RAP unit of Kisatchie NF (June–Aug). In WFE, it is uncommon along forest edges with sightings in late Apr, then June to Oct. In the FL Parishes, records in EBR were from May, Aug, and Sept, and at Asphodel Plantation from Apr to Nov. Brou had numerous specimens in his collection, from his yard near Abita Springs, as well as from Greensburg.

Host Plants: In southeast TX, it is reported to use purpletop tridens, also known as greasegrass, and probably others.

Arogos Skipper (*Atrytone arogos*)

ventral

dorsal male

dorsal female

Arogos Skipper *(Atrytone arogos)* *(continued)*

ventral

Description and Behavior: The Arogos Skipper has an eastern and prairie subspecies. Per Schweitzer et al. (2011), the eastern subspecies, *A. a. arogos*, includes a variant found along the Gulf Coast. In that region there are three broods, mid-Apr into May, June through July, and Aug to mid-Sept, most common during the latter period. From 1994 to 1998, the US Fish and Wildlife Service funded surveys by Minno and Minno (2006) to find populations of this skipper in the eastern portion of its range, including east LA and south MS. Unlike the other populations, in the Gulf Coast population the females were nearly black on the upper sides of the forewings. On the males, the veins in the yellow patch on the upper sides of the forewings were partly outlined in black.

My Records: During Aug 2012, I found a colony at Crosby Arboretum in Picayune, MS, located four miles north of the LA border. The skippers were in two pitcher-plant bogs. In mid-Aug 2013, I found it to be numerous in the pitcher-

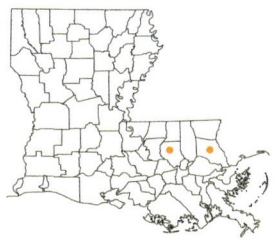

plant bogs at Abita Creek Flatwoods Preserve, about twenty-two miles as a crow flies from Crosby. I saw thirty, with a ratio of about three or four males to each female. The males were perched on tall yellow flowers (possibly honeycomb aster) within the bogs. One male spent the entire 2.5 hours I was present perched on the same flower. They seemed to stake out a flower as a territorial perch but would leave if another skipper landed on the same flower. While some of the females were found flying down in the grass, others were perched up on the tall yellow flowers like the males. Thirteen days later, eleven were recorded, again at Abita Creek Flatwoods Preserve. In early Aug 2014, a healthy colony was found at Lake Ramsey Savannah WMA, and then, in early Aug 2015, another large colony was located at Talisheek Pine Wetlands Preserve, both in STA. My only non-Aug record was a lone male at Lake Ramsey Savannah WMA in late May 2015.

Distribution and Abundance: This skipper was listed in 2015 as a Tier II, S1-ranked SGCN in LA. It was first found in LA in STA, in mid-Sept. Strickland (1972) found a large colony within the same parish in mid-Aug in open pine flatland, north of Lake Pontchartrain. He noted that as many as four could be found feeding on a single flower. Brou had two specimens in his collection from his yard (near Abita Creek Flatwoods Preserve)—a male caught in late Apr, and a female caught in early June. Via personal conversations with Strickland, he reported also seeing this skipper at Tickfaw SP in Aug during the early 2000s. In Aug of 2011 and 2012, I searched Tickfaw SP on three occasions without success.

Host Plants: In FL it is reported to feed on toothache grass. In the FL Panhandle, it has also been found to oviposit on broomsedge bluestem and lopsided Indian grass. Roever believes it uses little bluestem (personal conversations).

SKIPPERS

Delaware Skipper *(Anatrytone logan)*

ventral

dorsal male

dorsal showing dark scaling

Description and Behavior: The Delaware Skipper ranges over the southeast US with a continuous flight from May to Oct over three to four broods. The range maps within several national and regional guides exclude the LA Gulf Coast; however, as the records below reveal, this skipper is present all along the coast. While never particularly common, neither is it rare, but present in multiple types of habitat. Bright yellow/orange dorsally, it is a medium-sized skipper

dorsal female with dark scaling

with the females larger than the males. Both sexes are avid flower visitors. The males perch and are wary, darting off with a swift flight, not always to return. The females have a more fluttery flight, fly lower, and seem to move away only a few feet before alighting again on a blade of grass.

My Records: I first saw this species in early Sept 2007, in a pipeline cut at Thistlethwaite WMA. Possessing extremely dark ventral shading, it was identified (with help) as a male Delaware Skipper. In mid-Aug 2009 and early May 2010, back at Thistlethwaite WMA in a low, heavily wooded area bisected by a slough, I found several of both sexes in a section of lush knee-to-waist-high broad-leafed grass. All specimens shared the dark scaling noted above, suggesting that dark scaling was not tied to season. Since then, I have also found it at the Acadiana Nature Walk, Indian Bayou WMA, Bayou Teche NWR, Freshwater City, Tickfaw SP, Abita Creek Flatwoods Preserve, Lake Ramsey Savannah WMA, and Talisheek Pine Wetlands Preserve. The first four were in low, damp, deciduous wooded habitat like Thistlethwaite WMA. All of the specimens from these lo-

Delaware Skipper *(Anatrytone logan)* *(continued)*

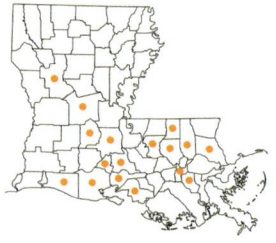

cations possessed the extremely dark ventral scaling initially noted. In contrast, I found a couple of much lighter-scaled males at Avery Island (June) and Duralde Prairie (Sept), and then more of the lighter versions in CAM (July, Sept, and Oct). The lighter specimens all seem to live in open, prairie-like habitat, while the darker versions have been found in or near forest-dominated habitat. That said, the habitat at Abita Creek Flatwoods Preserve, Lake Ramsey Savannah WMA, and Talisheek Pine Wetlands Preserve is open pine flatwoods, yet the specimens there have been extremely dark.

Distribution and Abundance: Strickland (1972) was the first in LA to list this skipper, as *Atrytone delaware*, in July, near a partly drained swamp south of Baton Rouge. He collected more in May, June to July, and Sept in varied habitats, observing, "Specimens are generally larger and darker than those examined from more northern localities in the U.S." Brou had multiple specimens from central LA and the FL Parishes. South LA records include sightings along the Bayou des Allemands at the LAFO-SJB line (Aug). It has been reported on four NABA counts: Bonnet Carré (Aug), Cameron (Aug), Pearl River (June), and Global Wildlife (July and Aug).

Host Plants: It is a grass feeder, including redtop panicgrass, bluestem grasses, broomsedge, and beardgrass. Charles M. Allen (2004) identified the grass at Thistlethwaite WMA and Indian Creek Rec Area as savannah panicgrass, a perennial in the southeast US. Savannah panicgrass has not been specifically identified as a host plant for *logan*, but several sources have suggested that other grasses found in wet, marsh habitat could also serve as a host plant. This skipper was reported to be on cane during the 2007 Pearl River count. I have often found it in areas of cane at Thistlethwaite WMA and Tickfaw SP, but do not believe this skipper uses cane as a host plant. Rather, I believe the connection has more to do with habitat preference.

Byssus Skipper *(Problema byssus)*

dorsal male

ventral male

dorsal female

Description and Behavior: The Byssus Skipper is another skipper with eastern and midwestern subspecies. The eastern subspecies's range is scattered, local, and generally rare across AL and MS, inhabiting moist or wet areas with tall grass. This subspecies generally flies from Apr or May to June and then Aug to Oct. The midwestern subspecies is uncommon to common across AR. Survey results from Rick Evans/Grandview Prairie WMA and Stone Road Glade Natural Area suggest two broods, late May into early June and then late Aug, found in wet areas with tall grass within blackland-prairie habitat. The males perch (on

Byssus Skipper *(Problema byssus)* *(continued)*

ventral

tall grass stalks and, in one instance, on the top of a pine seedling) and patrol (flying fast and direct, low, along the grass tops). The female's flight is slower and more bouncy, primarily down in the middle of the host plant. I have not seen any at flowers, but it is reported to be an avid flower visitor, including coneflowers, milkweed, butterfly bush, and pickerelweed.

Both sexes are bright orange-yellow ventrally; however, they are dimorphic dorsally. Both possess black dorsal borders, but the male exhibits more orange. The males are smaller, and their forewings are more pointed. The females are darker with a more blocked forewing. As both possess pale patches in the middle of the ventral lower wing, this skipper might be confused with the Yehl Skipper. Also, the female Delaware Skipper is dorsally similar to the male Byssus Skipper; however, Delaware Skippers do not have the pale patch in the center of the ventral hindwing. All three fly together at Rick Evans/Grandview Prairie WMA and Stone Road Glade Natural Area in close proximity to each other, so identification can be difficult at times.

My Records: My experience with Byssus Skippers has been exclusively in southwestern AR in late May, early June, and late Aug. I once found it numerous in June at Rick Evans/Grandview Prairie WMA, flying in an open area of thick, blooming Eastern gamma grass near the edges of moist woods.

Distribution and Abundance: The first record was by Cunningham (2011) in early June 2011, who reported a male in a slightly disturbed area near Chemin-A-Haut SP. Although it closely resembled a female Delaware Skipper, based on size, a slight difference in coloration from *logan*, and the complete border on the forewing, he concluded it was a *byssus* male, later confirmed by Andrew Warren. Roever has verbally reported finding this skipper in late May/early June at Sicily Island Hills WMA in areas where he found its food plant, gamma grass.

Host Plants: Cech and Tudor (2005) identified gamma grass, a tall, sharp-edged grass that grows in thick clumps, as a host plant. This grass is infrequent in the northern portions of LA, found in wet and/or moist areas. Indian grass has also been suggested as a possible larval food plant.

Zabulon Skipper (*Poanes zabulon*)

ventral female

dorsal female

ventral male

dorsal male

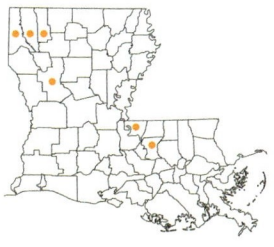

Description and Behavior: This is a woodlands skipper, found inside open deciduous woods, near water, and along the forest's edges, trails, and roads. The males perch on sun-splotched leaves, particularly late in the afternoon, when they can be found chasing after each other, darting in and out of the shadows. The females are much more sedate and reclusive. They will visit flowers such as blue asters that are in the sun along the forest's edge, particularly in the late afternoon. As the pictures reflect, this skipper is strongly dimorphic, with the males smaller and yellow with spots while the female resembles in both size and coloring a female Clouded Skipper.

My Records: This is another skipper I have not yet found in LA. The closest I have found it is in Rick Evans/Grandview Prairie WMA in AR.

Distribution and Abundance: Strickland first reported this skipper in LA at Greenwood Park in mid-Apr. It has since been recorded in north LA, central LA, and the FL Parishes. North LA records include Kisatchie NF in WEB (Apr), Eddie Jones Park (Apr, Aug, and Sept), and Bossier City (June). It was counted as part of the 2010 Shreveport count (June). It has been recorded in Kisatchie NF in NAT (Mar and Aug). Israel (1981) reported it to be uncommon in WFE with sightings near Bains (Apr) and Weyanoke (June to Oct). He described its preferred habitat and habits as "found on flowers, at sunny spots in dense thickets or along sides of forest trails." These records, plus my AR sightings, suggest three broods, in the spring, midsummer, and fall.

Host Plants: On the East Coast, it is reported to use numerous grasses such as bent grass, bluegrass, lovegrass, and purpletop.

Broad-winged Skipper *(Poanes viator)*

ventral

dorsal

Description and Behavior: This is one of LA's larger orange skippers. Primarily found along roadsides and/or pathways near water where its host plant is growing, it readily visits several types of flowers. Once disturbed, this skipper flies ("weaves" is the more accurate word) in and out of its host plant, making it difficult to follow. Markings on the ventral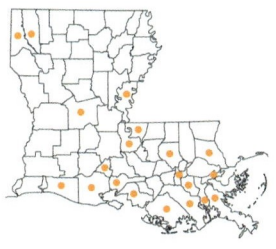
hindwing are somewhat similar to the Yehl Skipper's markings, but the two can be distinguished by the number of spots (four on the Broad-winged, three on the Yehl) and location (the Broad-winged is much more inclined to be around water).

My Records: In Shreveport, I found it at low-growing blue flowers along the edge of a small lake at C. Bickham Dickson Park. Along the Mississippi River corridor, I found two at the boat launch at Bayou Cocodrie NWR in CON. In the coastal region, I've seen it throughout CAM, where it can be widespread and common at tall growing asters along roadside ditches and canals, particularly in Sept and Oct. On one occasion in east CAM in Oct, it was abundant at every stand of blooming asters. I have also found it in ditches along the roadways south and east of Pecan Island in VRM (Oct), and then in a band along the IBE-SMA coast, south of Hwy 90, from Jefferson Island through Avery Island and Cypremort Point to Burns Point (Apr, July, Sept, and Nov). My latest sightings have been Nov 11 (Lacassine NWR) and Nov 12 (Rip Van Winkle Garden). I have found it within the city of Lafayette (more than once), and at Northlake Nature Center (mid-Aug). In the FL Parishes, I found a large colony at Big Branch Marsh NWR in STA (Sept).

Distribution and Abundance: It is locally common, ranging into all of the state's regions. In north LA, it is uncommon to common in CAD (C. Bickham Dickson Park, Cross Lake) with sightings between early May and late Oct. It has also been recorded at Bodcau WMA (Aug). Strickland (1972) had records "from LIV in the southeast to CAM in the southwest to PCP near the southwestern corner of MS." His records were from Mar to Oct, excluding July, and its habitat was restricted to marsh, edges of streams, and other bodies of water. There are multiple records from the NO region (Bayou Sauvage NWR, June) and the FL Parishes (Mandeville, Oct). It has appeared on the Barataria count regularly between 2004 and 2007 (June and July), once on a New Orleans count (June 2005), and once at Bonnet Carré (Aug 2013). Other southern parishes include

Broad-winged Skipper (*Poanes viator*) (*continued*)

SCH (Mar), TER (Oct), and LAFO. The combined records suggest at least four broods, Apr–May, June, July–Aug, and then Sept–Oct.

Host Plants: In CAD, this butterfly has been found mainly at the edges of water in and around giant cutgrass, its host plant. This grass grows from six to nine feet tall with blades about one inch wide. Strickland identified the same host plant, referring to it as marsh millet. It was reported as associated with sedges during the 2013 Bonnet Carré count. In southeast TX, it is reported to use common reed, wild rice, and sedges.

Yehl Skipper *(Poanes yehl)*

ventral female

dorsal female

dorsal male

ventral male

SKIPPERS

Yehl Skipper *(Poanes yehl)* *(continued)*

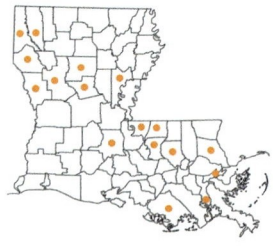

Description and Behavior: This skipper is one of four cane-feeding butterflies/skippers that are regularly found in LA. It is easy to approach when at flowers but can disappear quickly once disturbed. Glassberg (2012) depicted a female with three distinct spots on the ventral hindwing but without a "pale ray" extending from the middle cell and between the spots. In my experience in LA, both males and females possess that ray, although in some instances the ray is very pale. Records across the state suggest two broods, May–June and Sept–Oct.

My Records: I first found the Yehl Skipper at Thistlethwaite WMA, where it is uncommon throughout the season. I found it to be common in late Sept at Tickfaw SP, with both sexes seen at wildflowers, including blue mistflower, Brazilian vervain, and a tall, weedlike plant that generated white flower heads like mistflower. It was also present at Sicily Island Hills WMA in early June at purple beebalm, very common one year but in significantly reduced numbers the next. It can be abundant at Abita Creek Flatwoods Preserve in late Sept. In all of these locations, it was found in areas near cane.

Distribution and Abundance: Its range is restricted by the limited range of its host plant, but it can be locally common when present. In north LA, it is uncommon in CAD, from mid-May to mid-June and then from early Sept to mid-Oct. Other north LA locations include Bodcau WMA (May and Sept), the Stonewall area (Sept), the Many area (June), and at Kisatchie NF in WIN (Sept and Oct). In the central LA region, records include Kisatchie NF in NAT (Aug to Oct) and Catahoula Butterfly Garden (Oct). It has been described as locally common in the FL Parishes from May through July and Sept through Oct, typically found at the edges of wooded areas where switch cane was present. Waddill Wildlife Refuge, Greenwood Park, and Northlake Nature Center (mid-Oct) are specific locations. It has been reported on several NABA counts in the southeastern region: Barataria (June), Pearl River (June on multiple counts, Sept in 2013), and New Orleans (June, two counts).

Host Plants: As indicated above, switch cane and giant cane.

Aaron's Skipper *(Poanes aaroni)*

ventral

dorsal

SKIPPERS

Aaron's Skipper *(Poanes aaroni)* *(continued)*

Description and Behavior: This skipper is best searched for at flowers. It seems to be particularly attracted to goldenrod and Brazilian vervain. One May, I saw three on showy evening primrose, a flower that typically attracts very little attention. At Sabine NWR, it can be found on low-growing white asters blooming out in the brackish marsh. Its larger

size and dark-orange coloring make it distinctive and easy to locate. The subspecies in CAM is *P. a. bordeloni*, which is sympatric with Bay Skippers in that region. Bay Skippers are also large orange skippers that often visit the same flowers, so there is the potential for confusion; however, upon close inspection, the "rays" on the ventral hindwings of Aaron's Skippers should have two dots bracketing that ray. A few of the specimens found in VRM had a second dot below the ray (for a total of three), while several had no dots at all below the ray. Bay Skippers are a brighter yellowish tan color and will not have any dots, either above or below the ray. Palatka Skippers, also present but much rarer, will not have either the ray or the dots.

My Records: I did not record it until 2010, but have since seen it from May through early Nov throughout CAM. It is a regular at Sabine NWR (at the Blue Goose and West Cove trails). It can also be seen in the roadside ditches along Hwy 82 in extreme southeast CAM. This skipper is also present in the Freshwater Bayou Lock Rec Area in Oct, where it can abundant. I believe there are as many as three broods, May, July–Aug, and Sept–Oct.

Distribution and Abundance: Strickland (1972) reported taking a Palatka Skipper from CAM; however, the specimen in his collection is an Aaron's Skipper, the first recorded within the state. This skipper has been reported by Gatrelle (2000) in southwestern MS (late Sept and Oct), so this skipper could also be found in southeast LA along the Gulf Coast.

Host Plants: On the East Coast, it is reported as possibly using smooth cordgrass and saltgrass. In FL, maidencane is listed.

Palatka Skipper *(Euphyes pilatka)*

ventral

dorsal male

ventral female

Palatka Skipper (*Euphyes pilatka*)   (continued)

Description and Behavior: The subspecies in LA is *E. p. pilatka*, whose range is along the Atlantic Coast into the FL peninsula, then west across the Gulf Coast into LA. It was listed in 2015 as a Tier II, S1-ranked SGCN in LA. Reported as a large skipper in FL, similar in size to the Brazilian Skipper, those that I have seen in LA and MS are not that large, but are more similar in size to Broad-winged Skippers. The sexes are dorsally dimorphic, and some females have a small pale patch in the middle of the lower ventral wing. Sources suggest that males perch on or near the host plant, and that these skippers will gather in numbers at nectar sources. In CAM, I have found it only at flowers, particularly goldenrod and Brazilian vervain, and never in appreciable numbers. In most instances, it is wary and hard to approach. In extreme southeastern MS and STA, I found it to be less wary while at blooming pickerelweed and liatris, but, once spooked, it rapidly disappears into the marsh, where bipeds cannot follow. In the southern portion of its range, it has two, possibly three broods, flying from May into July and then from Aug into Oct.

My Records: I have seen it at Cameron Prairie NWR and Sabine NWR in Oct, flying with Bay and Aaron's skippers. In late Sept/early Oct 2016, I located a large colony with both males and females at Big Branch Marsh NWR (STA) along a roadside ditch in an open-marsh section of that NWR.

Distribution and Abundance: Roever verbally reported finding it at Avery Island in early June at the bird rookery. Rosemary Seidler has verbally reported seeing it in CAM.

Host Plant: On the East Coast, it is reported to use only sawgrass, a sharp-edged sedge that grows in very limited environments.

Dion Skipper *(Euphyes dion)*

ventral male

dorsal male

dorsal female

Dion Skipper *(Euphyes dion)* *(continued)*

ventral

Description and Behavior: This skipper has a widespread distribution but is uncommon within that range. Across its general range, it is reported to have two to three broods and is found in freshwater marshes, wetlands, swamps, and bogs. The male perches, but I've also seen it appear to be patrolling. It will visit flowers, but less so than Dukes' in my limited experience. I have seen it at pickerelweed and liatris, both blue flowers, in Sept. The Dion Skipper is about the same size as the Dukes' Skipper, and the two species can be difficult to distinguish. Shuey (1989) noted that southern Dions were "consistently dark." The males show much more tan dorsally than either of the Dukes' sexes. The Dion female has a tan-colored bar against a dark background on the dorsal hindwing that the Dukes' female lacks.

My Records: In Aug 2010, I found a small colony of this skipper on consecutive weekends in a roadside ditch along the main road through the center of Thistlethwaite WMA. None were found in early Sept after the ditch had been mowed. After a dry summer in 2011, I returned in Aug but found no Dions in

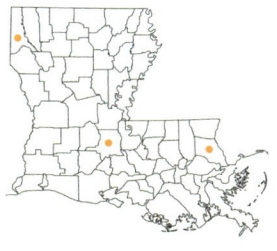

that ditch (which, again, had been mowed), but in a pipeline cut where there has always been a very vibrant colony of Dukes' Skippers, I found five to six Dions, including a mating couple. A week later, I found twenty-two Dukes', but no Dions in that pipeline cut. In Sept 2013, I found a male at Abita Creek Flatwoods Preserve on low-growing blue flowers, which I suspect were a member of the ageratum family. These flowers were extremely popular, attracting numerous different species of skippers. In early Sept 2017, I saw four feeding at pickerelweed in Big Branch Marsh NWR.

Distribution and Abundance: The first report of this skipper in LA was by Strickland, taken in CAM; however, I have determined those specimens were Bay Skippers, not Dion Skippers. The only other report from LA (outside of my records) was verbal by Roever from the north shore of Lake Wallace in CAD (June). Because it was reported in southwest AR and east TX, I expected it to be present in northwest LA, and in 2017 Patterson reported by email finding it near Tallulah (MAD) in early Sept 1985 (this late report is not reflected in the distribution map). Reported by Shuey (1989) along the MS Gulf Coast, it should also be present in the FL Parishes, south to the NO region and then farther south. It was listed in 2015 as a Tier II, SU-ranked SGCN in LA.

Host Plants: This skipper is reported to use various sedges (such as shoreline sedge, hairy sedge, and woolgrass, a freshwater sedge) and rushes.

SKIPPERS

Bay Skipper (*Euphyes bayensis*)

ventral

dorsal female

dorsal male

Description and Behavior: In 1989, Shuey identified *E. bayensis* as a new species "based on morphological and limited biological evidence." He differentiated Bay Skippers from Dions based on color, pattern, and habitat with the former having an expanded orange pattern, particularly on the male dorsally,

and Dion Skippers being much darker. Further, Bays frequented brackish marsh while Dions normally occurred in freshwater wetlands (Shuey 1989). While at flowers, Bays are docile and less wary than Dion Skippers. They might be confused with Aaron's Skippers; however, the "rays" on the ventral hindwings of Bays lack the two dots that should bracket the ray and would be diagnostic of Aaron's Skippers.

My Records: I first found this species in mid-Oct 2010, in west CAM around Holly Beach, flying with Aaron's Skippers. Their identification was confirmed by Andrew Warren. Those first specimens were seen at stands of blooming goldenrod in brackish marsh. Later, during that same month and into early Nov, I found both sexes across CAM to its eastern border, including Sabine NWR, Cameron Prairie NWR, and near Rockefeller NWR, at goldenrod, Brazilian vervain, tall asters, frog fruit, and on a low-growing, matlike vine which I believe to be a member of the pea family and which produced a small, multi-leaved yellow flower. In east CAM, it was flying with Aaron's and Broad-winged Skippers. In Oct 2013 and 2014, I extended my search into west VRM near Freshwater Bayou Lock Rec Area, where I found it again flying with Aaron's Skippers (albeit in fewer numbers) at low-growing blue asters and goldenrod. One, a male, had just been captured by a yellow crab spider on goldenrod. My records include May and July through Nov. I believe there are at least three broods—spring, midsummer, and late fall.

Distribution and Abundance: I was not the first to find Bay Skippers in LA; rather, Strickland's (1972) CAM specimens, taken near the courthouse in the community of Cameron during 1968 and 1969 and described by him as *E. dion alabamae*, were actually Bay Skippers. I was able to make contact with him, and he was kind enough to allow me to study his specimens, a male and a female, which matched those pictured in Shuey's 1989 article. The Bay Skipper was listed in 2015 as a Tier I, S1-ranked SGCN in LA.

Host Plants: After visiting some locations in CAM that I had identified for him, Salvato (2011) suspected salt-marsh bulrush after witnessing this skipper in close association with that saline-associated grass. He also identified sawgrass and reeds as potential larval host plants.

Dukes' Skipper (*Euphyes dukesi*)

ventral

dorsal male

dorsal female

Description and Behavior: This is a large, dark skipper about the size of the Broad-winged Skipper. The males patrol with a slow, somewhat bobbing flight just above and through the sedge food plant. The males will occasionally perch. The females fly low, among and through the tall sedges that dominate their habitat. Both sexes readily come to flowers, usually white and blue flowers, where

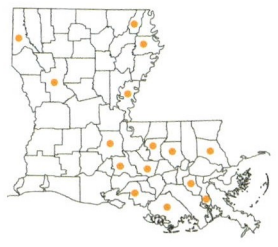

they are approachable. They have a swift flight when disturbed. Continuously on the wing from Apr to Oct, the Dukes' Skipper ebbs and flows in number over, I believe, three broods—spring, midsummer, and fall. It is similar in appearance to the Dion Skipper, as both have a light ray down the middle of the ventral hindwing; however, the dorsal wings of the Dukes' are much darker (almost all dark brown) and the forewing shape is wider and less pointed.

My Records: Typically, this skipper resides in habitat that is wet and heavily overgrown, with many mosquitoes, snakes, and even alligators. In Acadiana, I have found it at Thistlethwaite WMA, Bayou Teche NWR, and Indian Bayou WMA, all near Carex sedges, around which the females can be found. I witnessed a female ovipositing on Carex sedge at Bayou Cocodrie NWR in CON. At Thistlethwaite WMA, this skipper can be found out in an open pipeline cut, while at the other two locations it stays back in the trees or along the tree line. In the fall of 2010 at Thistlethwaite WMA, it was present throughout the WMA where ironweed was blooming. Jeff Trahan and I found one in the NAT unit of Kisatchie NF, in an area that was not typical for this skipper with no sedge in sight. It has not been recorded there since.

Distribution and Abundance: This skipper is more prevalent in LA than the field guides would suggest. Strickland (1972) described it as common in the proper habitat, which is essentially wherever its host plant is found. He had records from north LA and the FL Parishes (with over one hundred specimens in EBR). While most often found in swampy habitat, he noted it was even found in EBR along a railroad track with no water and little shade. His records were from May through Nov. More recent north LA records include Tensas River NWR (Aug). Other records include STA, IBV (St. Gabriel), and lower TER (May and Oct). It has been documented regularly on the Bonnet Carré count, with records from Aug on eight counts and once on the 2015 Pearl River count.

Host Plant: Strickland (1972) reported it as feeding on shoreline sedge, which grows two to three feet tall in wetter areas. Except for the referenced incident at Kisatchie NF, I have always found it associated with sedges.

Dun Skipper *(Euphyes vestris)*

ventral

dorsal male

dorsal female

ventral

Description and Behavior: This is the quintessential "little brown skipper." The males are chocolate
brown with very little, if any, markings, while the
females have white spots on their forewings, both
dorsally and ventrally. Dun males are smaller than
Dun females. The male's primary distinctive feature
is a gold-colored head; however, this coloring fades

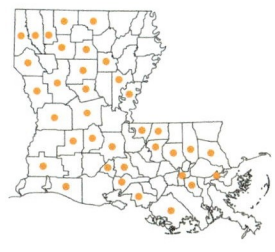

with age. Duns are most often found in open grassy areas along roads, ditches,
and pipeline or power-line cuts. This skipper flies low, in and around the grass.
Its typical flight is leisurely, moving in a bouncing manner. It is fond of many
kinds of low-growing flowers, where it can be approached closely. The males
will also gather at dung, mud puddles, and wet spots in dirt roads; however,
while engaged in this behavior, they are quite wary and hard to approach. The
Dun Skipper is similar in appearance and habits to the Clouded Skipper, with
which it sometimes flies. Female Duns are typically smaller than Clouded
Skippers and show no white frosting or clouding, which is usually exhibited
ventrally on the Clouded.

My Records: I have found it at numerous locations around the state, flying
from Apr until Oct. I believe there are at least three broods, spring (Apr to
May), midsummer (June to July), and fall (Aug to Oct). There may be a fourth
brood in late fall that accounts for those seen flying in Oct. I see it commonly
at Thistlethwaite WMA and regularly in the Acadiana region.

Distribution and Abundance: Lambremont (1954) first reported this skipper (as
Atryone ruricola metacomet) but, despite it being a relatively common skipper, it
was not reported again until after 2000. In CAD, it is uncommon to common
from early Apr to the end of Oct. In central LA, it has been reported in RAP
(Aug–Sept) and SAB (May). It is also present in the Felicianas from Apr to Oct,
described as uncommon. It has also shown up in several NABA counts such as
Cameron (July), Global Wildlife (July), New Orleans (June 2003 and 2005),
Pearl River (June), Shreveport (June and Sept), and Catahoula (Sept). There
was one lone individual reported for the 1999 Bonnet Carré count, with none
before or after.

Host Plant: In southeast TX, it is reported to use sedges.

SKIPPERS

Dusted Skipper *(Atrytonopsis hianna)*

ventral

ventral

ventral

Description and Behavior: The Dusted Skipper is primarily chocolate brown with a "dusting" of white scales along the outer margins of the forewings and hindwings. Cech and Tudor (2005) described the species as "taxonomically complex" with two very different-looking subspecies/races, northern *hianna* and

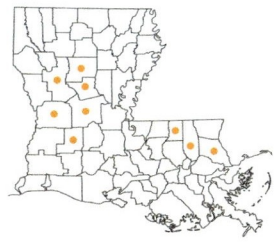

southeastern *loammi,* which some consider to be "sibling species." The principal difference between *hianna* and *loammi* is the number of white, opaque spots, primarily on the hindwing. The skipper is described as fairly large, especially *loammi.* Typical FL *loammi* are double-brooded, whereas all *hianna* are reported to be single-brooded. It is unclear which subspecies is present in LA, but, as a species, it was listed in 2015 as a Tier II, S3-ranked SGCN in LA.

My Records: I first saw this skipper in late Mar 2009, at the NAT unit of Kisatchie NF at one of the Caroline Dormon trailheads. Seen in that unit on several subsequent occasions, most have been found at flowers along roads, while others were found in open areas with Little Bluestem. The latest I've seen this skipper was the third week of Apr, at the NAT unit. The males were typically *loammi*-like, while the females were more like *hianna* in appearance. The females were primarily seen at flowers (yellow coreopsis, low-growing blue phlox, thistle, and indigo), and the males were perched on dead grass or at bare spots on the ground. In Apr 2010, at the RAP Kisatchie unit, I found several strongly marked *loammi,* but one week later in NAT, the markings on those found were much reduced. I had similar experiences in the spring of 2011 and 2013.

Distribution and Abundance: This skipper has been recorded only in central LA and the FL Parishes. Strickland (1972) first found it, in SHE and NAT, primarily in Mar and Apr, but also Aug. The area in SHE was a tree farm that used controlled burning, thereby creating a disturbed, open, hilly longleaf pine forest with a "heavy cover of grasses . . . and many flowering plants." Within this habitat, Strickland found numerous specimens between March 22 and April 8. The males patrolled at times and at other times defended territories. Their flights were between two to four feet above the ground. Strickland took one specimen on a common field grass in the bluestem family. He had a total of four specimens from early Aug, all in SHE, but he was not convinced that these represented "a reproductively successful second flight." Brou had specimens in his collection from near Fluker, taken in mid-Apr, early Aug, and mid-Sept.

I had an opportunity to examine the Strickland and Brou collections. The series from SHE varied from "almost immaculate dark specimens to those with broad, heavy white areas (Strickland 1972). Some of Brou's fall specimens were

Dusted Skipper (*Atrytonopsis hianna*)   (*continued*)

ventral

lightly spotted, with one bearing very little spotting. Roever, who collected a series of seventy-six of this skipper with Strickland, verbally reported the trend of males appearing more *hianni*-like, while females in the same area possessed *loammi* tendencies. Photographs of specimens provided to me by Seidler and Trahan were all in Mar and Apr. Some of their specimens were clearly *loammi*-like. Other individuals from WIN (Apr) and NAT (Mar) were not strongly marked. Recently, Roever verbally reported catching a *hianna*-like specimen at Lake Ramsey Savannah WMA, and David Patton sent me pictures of a *loammi*-like specimen at Dove Field Rec Area, both in early Apr.

Host Plants: In the East, this skipper is reported to use big and little bluestem.

Pepper and Salt Roadside-Skipper (*Amblyscirtes hegon*)

ventral

dorsal

ventral (alternate coloring)

SKIPPERS

Pepper and Salt Roadside-Skipper *(Amblyscirtes hegon)* *(continued)*

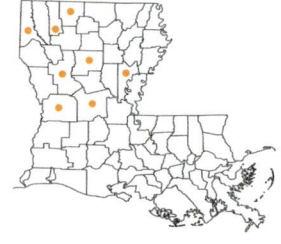

Description and Behavior: This is a small, early spring skipper, flying only in Mar and Apr. Like the other roadside-skippers, it flies low to the ground, regularly stopping at various low-growing flowers. On the ventral hindwing, the Pepper and Salt Roadside-Skipper will have some light, cream-colored spots in the basal cell and then a cream-colored postmedian band. At Sicily Island Hills WMA, I would estimate that one out of five seen were tan colored dorsally rather than the typical slate gray (see photos). I did not find any source-material explanation of the tan-colored form, but several sources had pictures of tan-colored specimens from other states. All of LA's roadside-skippers have checkered wing fringes, so there is some potential for confusion. This skipper can be distinguished from the Lace-winged by the absence of any cream-colored veins on the ventral hindwing.

My Records: In the NAT unit of Kisatchie NF, I have found it on a trail that leads down to Kisatchie Bayou in early Mar. I have also found a few in early Apr in a power-line cut that crosses Castor Plunge Road in the RAP unit. In both locations, I played an extended game of cat-and-mouse, trying to get close to individuals perched on the ground. I have also found it in the WIN and VER units of Kisatchie. This skipper was abundant (Mar) along several gravel roads through the woods at Sicily Island Hills WMA. In the coolness of the morning, they were basking in the roads. Later, they were found at wild garlic. While wary on the road, they were more approachable at flowers. The lesson learned is to wait for this skipper to go to flowers.

Distribution and Abundance: This skipper was listed in 2015 as a Tier III, SU-ranked SGCN in LA. It is uncommon in CAD (Eddie Jones Park) from mid-Mar to mid-Apr. It has also been found in Kisatchie NF in WEB (Apr) and CLA (also Apr). Strickland (1972), calling it *A. samoset*, found it at Vowells Mill in NAT in early Apr. Strickland described the locality as the longleaf pine upland region of central LA. Other records from NAT include the Kisatchie unit along the Scenic Byway (Mar and Apr).

Host Plants: It uses various types of grasses, including river oats, fowl manna-grass, and Indian grasses.

Lace-winged Roadside-Skipper (*Amblyscirtes aesculapius*)

ventral

dorsal

SKIPPERS

Lace-winged Roadside-Skipper (*Amblyscirtes aesculapius*) (*continued*)

Description and Behavior: The Lace-winged Road-side-Skipper is another of the four cane-feeding skippers that can regularly be found in LA. This skipper flies in early spring and then again into late summer. Israel (1981) reported three broods (spring, midsummer, and fall). Although it was listed in 2015 as a Tier II, S3-ranked SGCN in LA, it is the most common

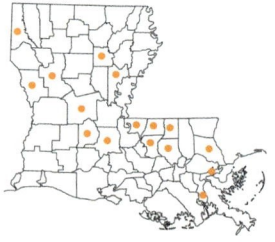

roadside-skipper in the state. It is also the most easily recognized, with its pale white cobweb-like pattern on its ventral hindwings. It readily comes to flowers, such as Brazilian vervain and frog fruit. During a June visit I made to Sicily Island Hills WMA, large numbers were seen at purple beebalm along the roadsides.

My Records: I have records from Mar to Sept, more common in Apr and May, and then again from Aug into Sept. I have found it at Copenhagen Hills (June and July), Chicot SP (July), and at Sicily Island Hills WMA (Mar and early June). It can be extremely numerous at Thistlethwaite WMA on occasion, with large numbers in Aug 2006. It is present at the back of the Mary Ann Brown Preserve in the creek bottoms at wet sand. It is also present just across the state line at Clark Creek Nature Area in MS. Farther east, it can be common at Tickfaw SP in the fall, and at Abita Creek Flatwoods Preserve in an open field near a boardwalk through a swampy area. In all of these locations, a significant amount of cane is present.

Distribution and Abundance: This roadside-skipper was first reported by Lambremont (1954) as *A. textor* (called the Wovenwinged Skipper). It is uncommon in CAD during mid-Mar to mid-May and mid-Sept to mid-Oct. Dutton had pictures of it on his website from the Many area in SAB. In the Acadiana region, several were reported at Chicot SP in Sept. In the FL Parishes, it has been recorded as locally common at Camp Istrouma (Sept) and Greenwood Park (May and Oct). Israel (1981) reported it at Asphodel Plantation from Mar to Oct. He also recorded it in WFE, Mar through May, uncommon during the same time frame, "taken at flowers along the edge of dense forest and along wooded trails." It has been found in the Abita Springs area numerous times during Mar and Apr and again from Aug into Oct. It was reported on the NAT Kisatchie count in Oct (once), the Pearl River count in June (six years) and Sept (one year), as well as the New Orleans count in June (twice).

Host Plant: In southeast TX, it is reported to use giant cane.

Common Roadside-Skipper *(Amblyscirtes vialis)*

ventral

dorsal

SKIPPERS

Common Roadside-Skipper *(Amblyscirtes vialis)* *(continued)*

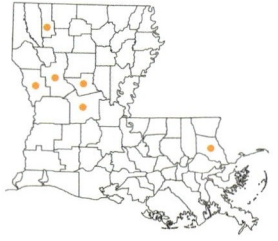

Description and Behavior: This roadside-skipper can easily be confused with the Dusky Roadside-Skipper, particularly since they can fly in the same location (such as the NAT unit of Kisatchie). There are some distinguishing markings, but, because this skipper and the Dusky are so small, seeing these markings in the field can be difficult. First, both skippers will show a checkered pattern along the fringes of their wings. On the Common Roadside-Skipper, that pattern is weakly expressed as alternating black and tan or light brown patches. Also, the antennae on this skipper are extended to a point at the tips.

My Records: I've seen this roadside-skipper during the spring in central LA. In the RAP Kisatchie unit, it was in a power-line cut along Castor Plunge Road. I saw another in the GRA unit along Stuart Lake Road. In both instances, they were found basking on a dirt four-wheeler trail.

Distribution and Abundance: This roadside-skipper is indigenous to the north and central LA regions. North LA records include CAD (Apr) and WEB (Aug). It has been recorded numerous times in the NAT Kisatchie unit (where it can be fairly common in the Red Dirt Area in Mar and Apr) and Vowells Mill (early Mar). Other records from that region include SAB and GRA (near Bentley). It was reported once in June, on the 2000 West STA count. (This might have been *A. alternata*.)

Host Plants: It is reported to use grasses such as bluegrass, Bermuda grass, bent grass, and inland oats.

Dusky Roadside-Skipper (*Amblyscirtes alternata*)

ventral

dorsal

Dusky Roadside-Skipper *(Amblyscirtes alternata)* *(continued)*

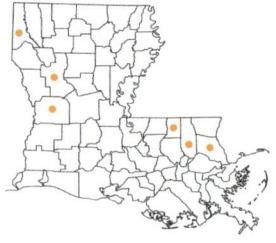

Description and Behavior: This species was listed in 2015 as a Tier II, S2/S3-ranked SGCN in LA. This is possibly the smallest roadside-skipper of the five found in LA. The males patrol with a flight that is swift and hard to follow. At Sandy Hollow WMA, they were found most easily at blooming false garlic. At the NAT unit of Kisatchie NF in mid-June, I found them along with Meske's Skippers at yellow flowers in a roadside ditch. The checkered patterns on the wing margins of this roadside-skipper are more boldly expressed than the Common Roadside-Skipper, in an alternating black-and-white pattern. Also, this skipper's antenna ends are more rounded and clublike, without a point.

My Records: Prior to 2017, I had found this skipper only in three locations, the VER and NAT Kisatchie units (Mar, Apr, and mid-June) and Sandy Hollow WMA (late Mar and Sept). I netted a specimen of this species in Kisatchie NF in RAP in mid-June of 2017 (this late record is not reflected in the distribution map). At Sandy Hollow WMA, I found it to be somewhat common in the spring with at least twelve and ten counted, respectively, flying along the side of the entry road in stands of dead little bluestem grass. It was much less numerous in Sept.

Distribution and Abundance: This small, innocuous skipper was initially found by Strickland (1972) in the longleaf pine–scrub oak hills near Greensburg, uncommon from late Mar into Apr and then again in June and July. It has since been reported in CAD, NAT, VER, and STA. At the NAT unit of Kisatchie NF, it was found flying with *hegon* and *vialis* in Mar and early June, along the Scenic Byway. Patton found one in June at Dove Field Rec Area. Brou had several specimens, collected in the FL Parishes in Apr and May, and again from July to Sept. One was reported during the 1977 Pearl River count.

Host Plant: It is reported to use bearded skeletongrass in FL, GA, and AL (Schweitzer, Minno, and Wagner 2011).

Celia's Roadside-Skipper *(Amblyscirtes celia)*

ventral

dorsal

SKIPPERS

Celia's Roadside-Skipper (*Amblyscirtes celia*) (*continued*)

Description and Behavior: This skipper's primary range is across TX, and it is uncommon in the Gulf Coast region. It has been identified as the most common roadside-skipper in Galveston County and the Houston area, flying from May to Sept. It is very similar to the Bell's Roadside-Skipper which, at one time, was considered a subspecies. In LA, this roadside-skipper does not fly in the same area as any of its cousins, so location alone should be a distinguishing factor. When fresh, this species will have some pale markings (not really spots) that are visible against the slate gray color of the ventral hindwing. It is reported to avidly feed at flowers.

My Records: I've seen it only at Barton's Creek in Austin, TX, where it was common in heavy woods along the creek, perched on and around several species of tall grass (including inland sea oats, which has been identified as a host plant in that area).

Distribution and Abundance: It has only been recorded in CAM, initially collected by Strickland during 1968–69, from Apr through Sept, with twenty-one seen in early June. The habitat was described as the shady areas within the cheniers along the Gulf Coast in the area of "East Jetty Woods" and Little Chenier. Strickland's (1972) specimens were confirmed to be this species by genital dissection. Despite two hurricane hits in the mid-2000s, Roever verbally reported finding this skipper in early Apr 2008, in the same area as reported by Strickland. This species was listed in 2015 as a Tier II, SU-ranked SGCN in LA.

Host Plants: In southeast TX, it is reported to use dallisgrasses, including St. Augustine.

Eufala Skipper *(Lerodea eufala)*

ventral

dorsal

SKIPPERS

Eufala Skipper (*Lerodea eufala*) (*continued*)

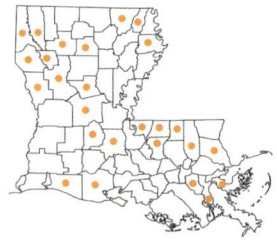

Description and Behavior: This is a small skipper, similar in size to the Swarthy Skipper but gray rather than brown, with white spots on the upper wing. Like the Swarthy, it flies low, in disturbed, open areas with grass. I've found its flight to be slower than the Swarthy's, and even when disturbed, it doesn't fly far before seeking refuge down in the surrounding vegetation, where it can be easily missed.

My Records: I find it most often at Thistlethwaite WMA. It appears in Aug in the ditches along the second road that crosses the pipeline cut as well as in that cut. I have seen it in the spring around Catahoula Butterfly Garden (late Mar into Apr). I have also found it at Indian Creek Rec Area. I have found it all over CAM, primarily in and around ditches along the roadside, from Aug to Nov. My only eastern record was at Talisheek Pine Wetlands Preserve (Aug).

Distribution and Abundance: Older records suggest this skipper was rare within the state, but it does not now appear to be as limited as suspected. It is common in CAD from early July into early Nov. It has also been found across the north and central LA regions at Bodcau WMA and Cane's Landing (Aug to Nov), the Red River NWR–Yates Tract (Sept and Oct), Kisatchie NF and Red River NWR–Lower Cane in NAT (Mar and Aug), Kisatchie NF in RAP (Aug and Sept), and Duralde and Eunice Cajun Prairie (Aug). In the Felicianas, it has been characterized as rare in Sept and Oct. Southeast LA records were in SBN, TAN, and JEF (all Sept and Oct). It has appeared regularly on several NABA counts, including Pearl River (June), Folsom (Oct with five reported), Shreveport, Catahoula (Sept), Global Wildlife (several times in June with a high of six in 2000), Cameron (July and Aug), and Bonnet Carré (Aug).

Host Plants: In southeast TX, it is reported to use numerous grasses such as Bermuda, St. Augustine, and Johnson grass.

Twin-spot Skipper *(Oligoria maculata)*

ventral

dorsal

SKIPPERS

Twin-spot Skipper (*Oligoria maculata*) (*continued*)

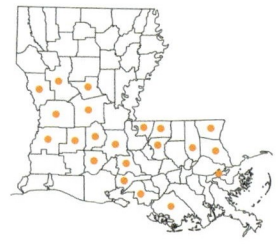

Description and Behavior: This skipper appears
to favor damp and/or wet, open areas near woods.
I have found it at Thistlethwaite WMA at flowers
alongside Dukes' Skippers, particularly on blue/
purple and white flowers. Never abundant and usually seen in low numbers, at Thistlethwaite WMA it
is much less common than Dukes' Skipper. I see it
most often at flowers, where it is easily approached. At Cooter's Bog, I found
several males perched on grass in and around the perimeter of a hillside pitcher-plant bog. In the field, it might be confused with both the Dusted Skipper
and the Clouded Skipper. The "*loammi*-like" Dusted Skipper is more of a chocolate brown color (as opposed to the Twin-spot, which is almost black) and will
have larger and more white spots on both the upper and lower ventral wings.
In contrast, the *hianna* Dusted Skipper will still have multiple, larger spots on
the forewings with reduced or absent spots on the lower wings. The Clouded
Skipper will have no white spots on the lower wing, either dorsally or ventrally.
The female Dun Skipper is similar in coloring but is both smaller and has no
white spots on the lower wings.

My Records: I have found Twin-spot Skippers more commonly in the Acadiana region than in central LA. I first found this skipper during the 2007
Thistlethwaite WMA count (June), just inside the tree line in a swampy area,
feeding on low-growing flowers. I've since seen it there many times, mostly in
the fall while it feeds on blooming ironweed. In Sept 2016, I found it in virtually identical habitat at Bayou Teche NWR. During mid-May, mid-June, and
late Aug, I found several specimens at Cooter's Bog and Leo's Bog. I saw one
at Talisheek Pine Wetlands Preserve in early Aug and several in a power-line
right-of-way near Enon in WAS in mid- to late Aug. In early Sept, I found one
in a wet section at Duralde Prairie and another at Sandy Hollow WMA. It was
common in STA at Big Branch Marsh NWR and Lake Ramsey Savannah WMA
in Sept 2017. I believe there are two broods, May–June and Aug–Sept, more
prevalent in the fall.

Distribution and Abundance: Records for this skipper are primarily from the
southern portions of the state. It was recorded during the Cajun Prairie survey

and has been regularly recorded in the central LA and Acadiana regions as well as the pine flats of the FL Parishes. Strickland found it to be common in EBR, taken in clearings in dense deciduous lowlands near marsh or swamp habitats, such as at Waddill Wildlife Refuge. During the Tunica Hills survey, this skipper was rare, found in May along the banks of Big Bayou Sara. At Asphodel Plantation, records are between May and Sept. It has also shown up on three NABA counts: Global Wildlife in July, Alexandria in Aug, and New Orleans in July. In the southernmost parishes, it has been found in the Schriever area (Sept) of TER.

Host Plants: It is reported to use bluestem grasses (Glassberg, Minno, and Calhoun 2000).

SKIPPERS

Brazilian Skipper *(Calpodes ethlius)*

ventral

dorsal

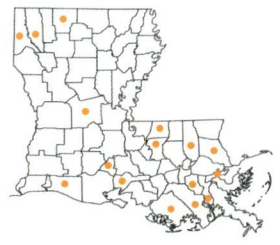

Description and Behavior: This large skipper has been reported regularly in thirteen parishes from all four corners of the state. The adult's flight is very rapid, rising and falling in a quick, undulating fashion. Brou reported verbally that this skipper remains active late into the evening after sunset. Cech and Tudor (2005) commented that it is easier to locate as a caterpillar than as an adult, and my experiences have matched their observation.

My Records: I saw my first Brazilian Skipper on Nov 1, 2013, at Rip Van Winkle Garden on Jefferson Island at about 4:15 p.m. Prior to a wedding, I saw a male feeding docilely at firespike in the warm afternoon sunshine. I have also since seen it in LAFA, once at a local nursery with ornamental canna for sale. I have found numerous caterpillars at stands of canna in both IBE and LAFA during the summers of 2014, 2015, and 2016, extending into Nov in 2016. The distinctive way that the caterpillars fold the canna leaves as a cover during daylight hours is very easy to identify. As the caterpillars grow, the size of the folded piece of leaf increases as well, making the caterpillar's presence even more apparent.

Distribution and Abundance: In north LA, there have been individual sightings in July, late Sept, and early Oct in CAD. In 2012, Trahan verbally reported raising several broods in Shreveport. In central LA, it was reported on the Alexandria count in Aug 2000 (with two seen), Sept 2013, and Oct 2016. In southwest LA, CAM records include July, Aug, Sept, and Oct (near Lacassine Pool and at Peveto Woods). In the FL Parishes, it has been found rarely at Asphodel Plantation during Aug to Oct. Brou verbally reported that at one time it was common at canna on his property near Abita Creek Flatwoods Preserve. It has been a regular in the NO area, including regular sightings during the New Orleans count in June 1996, 1998, and 2000, and July 1992, 1994, and 1995. It was reported on the Barataria count in June 2000. On the Bonnet Carré count, it was reported in Aug 1997, 1999, and 2013. Farther south, it is reported to be common on Grand Isle near canna in Apr, Aug, Oct, and even late Dec. Also in southernmost LA, it has been reported in the Houma area (Aug), at Galliano (mid-Dec), and at Golden Meadow (late Nov).

Brazilian Skipper *(Calpodes ethlius)* *(continued)*

Host Plants: In LA, as elsewhere, it uses all kinds of canna as its larval food plant. It was reported to be found on canna lily on the 1997 Bonnet Carré count (with nine noted) and then on yellow canna lilies on a Cameron count. In southeast TX, it is reported to use powdery thalia. In FL, it has been reported to use ginger, arrowroot, and alligator flag.

Salt Marsh Skipper *(Panoquina panoquin)*

ventral

dorsal

SKIPPERS

Salt Marsh Skipper (*Panoquina panoquin*) (*continued*)

Description and Behavior: Both this species and the
Obscure Skipper with which it often flies are smallish, rather nondescript skippers. They have a quick,
darting flight, low to the ground. Both are readily
attracted to flowers such as Brazilian vervain, asters,
goldenrod, frog fruit, and pickerelweed. They ex-
hibit a fluttery flight when moving between flowers.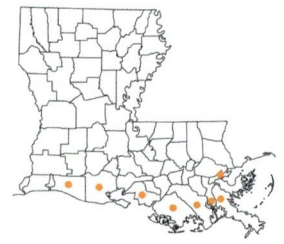
The following is a general but typically accurate method I have developed to
distinguish these two and Neamathla Skippers: Salt Marsh Skippers are usually
the largest and the least dark, showing more of a tannish brown rather than
a dark brown color. If approach is allowed, a white dash in the middle of the
ventral hindwing is diagnostic of a Salt Marsh. If there are two small white ver-
tically aligned spots on the ventral hindwing, it is an Obscure. If there is nothing
showing, and it is very dark, chances are good it is a Neamathla.

My Records: I had not seen this skipper before 2010, but that was because I was
not looking in the right places. It is present nearly everywhere near the coast in
CAM, with dates from May into Nov, common to abundant, particularly in the
fall. Locations include Sabine NWR, Cameron Prairie NWR, and Rockefeller
NWR. I also found it to be abundant in late Oct at Freshwater Bayou Lock Rec
Area. It is easily approached while at flowers.

Distribution and Abundance: This small skipper can be common on the Gulf
Coast in salt-marsh habitat, including the Chandeleur Islands at the mouth of
the Mississippi River, Port Fourchon, and Grand Isle. Dates were from Apr and
June to Oct. Other locations include SMA (May) and TER. It has also shown
up on two counts: Grand Isle (May 2003) and Barataria (June 2000), with
thirteen reported.

Host Plant: In southeast TX, it is reported to use seashore saltgrass.

Obscure Skipper *(Panoquina panoquinoides)*

ventral

dorsal

Obscure Skipper *(Panoquina panoquinoides)* *(continued)*

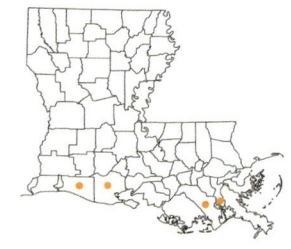

Description and Behavior: This skipper is typically less numerous than the Salt Marsh Skipper. It is smaller than the Salt Marsh (by just a bit), and the two often fly together. I've seen them both taking nectar from the same stand of Brazilian vervain and in the same patch of frog fruit. It also frequents the low-growing asters in the brackish marsh at Sabine NWR. Strickland (1972) commented that his specimens from southwestern LA had three spots on the ventral hindwing, while those from east LA had only one spot. All of the specimens I have found in CAM had two spots on the ventral hindwing.

My Records: I did not know this skipper until 2011 when I found it at Grand Chenier, near Rockefeller NWR, in Aug. It can also be found within Sabine NWR in May and Sept into late Oct. I also found it at several locations around the Holly Beach and Johnson Bayou areas in early Oct. In late Oct 2013 and 2014, it was present in low numbers at Freshwater Bayou Lock Rec Area.

Distribution and Abundance: Strickland first reported this skipper from CAM (June), on the Chandeleur Islands (Aug), and at Grand Terre (an island off the coast of JEF, Aug). Subsequent records include Port Fourchon (Apr, flying with Salt Marsh Skippers) and Broussard Beach. Recorded in only four parishes, it was listed in 2015 as a Tier II, S1-ranked SGCN in LA.

Host Plants: In southeast TX, it is reported to use Bermuda grass, sugarcane, and other coastal grasses.

Ocola Skipper *(Panoquina ocola)*

dorsal

ventral

Ocola Skipper (*Panoquina ocola*) (*continued*)

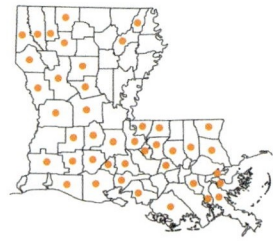

Description and Behavior: This skipper is common and can become abundant in the fall. It visits many different kinds of flowers, but remains wary and difficult to approach. It is a very fast flier, dashing around at speeds hard to follow. There are two colors forms, a lighter brown and a much darker gray-brown, which can be seen on the same day. I'm not sure if the colors are sex-based as it is extremely difficult to tell the two sexes apart in the field.

My Records: My records are primarily from June to mid-Nov (and even into Dec on occasion). In the fall, it can be one of the most common skippers on the wing, particularly in the southern portion of the state. It can be abundant at Thistlethwaite WMA in Aug and Sept, taking nectar at blooming ironweed and Brazilian vervain. It is very common in CAM in Oct and into Nov. In mid-Nov 2012, on a gray, windy, misty day, I easily saw five hundred along Streeter Road in Lacassine NWR, primarily taking nectar on low-growing yellow composite flowers in the roadside ditches. In 2012, a banner year, three were seen in LAFA as late as December 2.

Distribution and Abundance: In the north, it is common to abundant in CAD from early July to mid-Nov. In the east, it is present in both of the Felicianas, with records in Apr and from June to Dec when it is common.

Host Plants: It is typically listed as using grasses such as southern cutgrass and torpedo grass. Ross and Lambremont (1963) reported its use of cultivated rice in LA.

Yucca Giant-Skipper *(Megathymus yuccae)*

ventral

dorsal

Yucca Giant-Skipper *(Megathymus yuccae)* *(continued)*

Description and Behavior: The Yucca Giant-Skipper and Strecker's Giant-Skipper are unique in that, after hatching, the caterpillars generate a burrow and silken tube in the area of the terminal bud of the host plant within which they live. The larvae pupate within the burrow over the winter months. Beiriger (2003) reported the silken tubes are approximately five centimeters (two inches) in length. Eventually, the shape of the tube converts into a cone, apparently to facilitate emergence of the adult with a minimum of problems. There are some good pictures of these cones and tubes on BAMONA's web link for Strecker's. The tubes should be visible during the dead of winter while the surrounding undergrowth is dead, providing a guide for where to search for emerging adults during the flight period. The males patrol over yucca plants but are not territorial.

Male Yucca Giant-Skippers bask in the sun on the ground or on objects like fallen logs or stalks of grass, near yucca plants. When flushed, they dash off with a fast, bobbing flight, flying long circles before returning to virtually the same perch as previously used. On several occasions, I have watched them fly away in one direction, only to return a short time later from the opposite direction. They are quite distinctive in flight, flying at about head height with the white patches on their forewing visible as they fly. The females appear to show more yellow than white in flight. The females don't circle like the males, but fly low around yucca plants, landing to oviposit.

My Records: I have seen this skipper only in east TX.

Distribution and Abundance: This species was listed in 2015 as a Tier III, S1-ranked SGCN in LA and is very rare. It was first recorded by Ross and Lambremont in early Apr 1963, south of the MS state border near Bains in a "rugged hardwood upland region with many steep bluffs and deep ravines." Most of Strickland's (1972) records are from the FL Parishes, including SHE, Fluker, and Weyanoke. Roever reported by e-mail that he found larvae at yucca growing along the dirt road leading from Pond, MS, toward Woodville. This area is immediately north of the WFE line. In the central LA region, Roever verbally reported finding larvae, pupae, and adults at the NAT unit of Kisatchie NF in Mar in the early 1970s, but I have found no records since for that location. Roever also reported finding this skipper in the Kisatchie units of DES, CLA,

GRA, WIN, and BIEN parishes. His only "sightings"
within the last ten years were at Sicily Island Hills
WMA (one adult, tents and pupa) in late Mar 2013
and in mid-Mar 2017. There was a LA listserv report
(without photograph) from northern CAD.

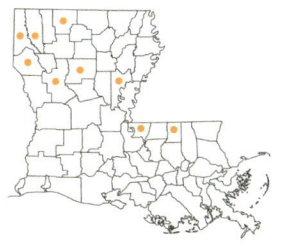

Host Plants: Strickland reported the food plant used
in SHE appeared to be Adam's Needle. In FL, this
skipper uses agave family members such as Spanish dagger, bear grass, Adam's
Needle, and Spanish Bayonet.

Strecker's Giant-Skipper *(Megathymus streckeri)*

ventral

dorsal

Description and Behavior: The females are extremely large with a swift flight. This skipper does not have a classic "butterfly-like" flight, and on one occasion, because of its size, dark coloring, and flight pattern, I initially thought I was seeing a dark cicada. I was actually able to hear the buzz of the female's wings in flight. When perched in the grass, with their wings folded over their body, these giant skippers become quite camouflaged; however, in flight, the yellow spots on the dorsal wings are evident. Roever has suggested this species is territorial and, when disturbed, will circle and then return, just as I have seen the Yucca Giant-Skipper do.

My Records: I have seen this unique skipper four times in the NAT unit of Kisatchie NF, all during the last ten days of Apr in different years. Three were females and were found in a specific area within the unit defined by a power-line cut with yucca plants in and around that cut. Two of the three specimens were basking in the sun along an old trail adjacent to that power-line cut. This trail is through an area of open pines with multiple yucca plants on either side.

Distribution and Abundance: This skipper was listed in 2015 as a Tier III, S1-ranked SGCN in LA. All LA records are from the Kisatchie unit in NAT, where it was first found in the early 1970s. Roever verbally advised that an LSU entomology professor, a Dr. Riley, "rediscovered" it there in the 1990s and that others have reported to him of taking this skipper in that same unit during the last ten years (as recently as Apr 2017, after a winter burn of the area). He took pupae there in early Apr 2008. Trahan located two and three in late Apr and early May 2013, respectively, some of which were females.

Host Plant: Spanish bayonet.

BUTTERFLIES

(Superfamily Papilionoidea)

SWALLOWTAILS (Papilionidae)

Pipevine Swallowtail (*Battus philenor*)

ventral

male dorsal

female dorsal

Description and Behavior: A common swallowtail in LA, the Pipevine Swallowtail can be found almost anywhere, from the piney woods in the north to the prairie marshes of CAM and all points in between. Males patrol by flying low to the ground in a swift, weaving flight. Although I have seen females ovipositing in open wooded areas, this swallowtail is not really a denizen of the deep, deciduous woods but prefers open, disturbed areas. It readily comes to flowers and can be drawn to city gardens, particularly if pipevine has been planted. It is very fond of mimosa blooms. While feeding, it hovers above the flower by using rapid, shallow wing beats. In early spring, the males can be quite small. On the average, males are smaller than females and show a much brighter, iridescent blue sheen on the dorsal lower wings. The females' lower dorsal wings are a much duller blue.

The Pipevine is reported to be unpalatable to birds due to toxins ingested during the larval stage. There are several LA butterflies that mimic it, all presenting an example of Batesian mimicry. These include the female black Tiger Swallowtail, the Red-spotted Purple, the female Black Swallowtail, and the female Spicebush Swallowtail, all of which

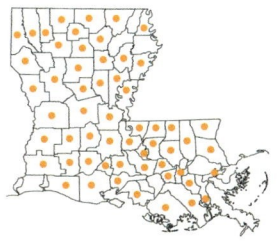

can regularly be found flying with Pipevines at Thistlethwaite WMA, Indian Bayou WMA, Avery Island, and at several Kisatchie NF units. As part of his Tunica Hills survey, Israel (1981) addressed this mimic relationship, providing frequency numbers for the model and its mimics. In that region, the Pipevine Swallowtail was generally not as common as the Spicebush Swallowtail or the Red-spotted Purple. The black Eastern Tiger and female Black swallowtails made up a small percentage of this mimic complex. Even during the time when Pipevines were most numerous, the mimics were more abundant.

My Records: I have seen it from Mar to early Nov, on both sides of the Mississippi River. It can be abundant at the NAT unit of Kisatchie in Mar. Another reliable location is the Catahoula Butterfly Garden. In late Sept 2013, I saw several fresh specimens flying in west CAM directly along the coast and in brackish marsh. All were quite large and suggestive of a late brood. It was still present at Peveto Woods in mid-Nov of that same year.

Distribution and Abundance: This swallowtail has a long flight season throughout the state, with an early recorded date of February 1 and late dates into Nov. It flies from late Mar to mid-Nov across north LA. Survey records from the Cajun Prairie region suggest it is the most common swallowtail in that region with sightings from Mar to Oct, most common in Mar, May, June, July, and Aug. In EFE, it is common, flying from Feb to Dec, most common in July and Sept. It has been recorded in the FL Parishes region and in southernmost LA (Apr to Aug).

Host Plants: In LA, it has been reported to use Dutchman's pipe or pipevines (native and cultivated). Snakeroot is reported as a common native host in FL. I have seen the caterpillars, which appear to be a centipede mimic with extended tubercles on both ends, while they were feeding on Dutchman's pipe. In fact, I noticed them only after I heard the readily audible sounds of two caterpillars chewing on mature leaves.

SWALLOWTAILS (Papilionidae)

Polydamas Swallowtail (*Battus polydamas*)

ventral

dorsal

Description and Behavior: This cousin of the Pipe-vine Swallowtail is the only swallowtail in North America completely devoid of tails. Its flight is similar to that of the Pipevine, but the yellow chevrons along the outer edges of both forewings are visible in flight and distinguishing. The typical range for this unique swallowtail is south FL and the Rio Grande

Valley. It has been reported in the Houston area and both the central and upper TX Gulf Coast regions as an occasional, rare visitor, so it might be possible in CAM and CALC.

My Records: In the continental US, I have seen it only in south FL.

Distribution and Abundance: There is an old sighting in ORL from 1863. The only other sighting of which I am aware was in mid-Aug 2016. John Himes, a biologist, reported on the LA listserv that he saw one at the Catahoula Butterfly Garden in GRA. Via e-mail, Himes advised he was familiar with this species through prior experience and was able to observe it at close range for several minutes while it fed at butterfly-bush blossoms. Despite the absence of a photograph, I was sufficiently satisfied with the veracity of this report to include it here.

Host Plants: In FL, it is reported to use members of the Dutchman's pipe or birthwort families such as calico flower, giant pipevine, and three-lobed pipevine.

SWALLOWTAILS (Papilionidae)

Zebra Swallowtail (*Eurytides marcellus*)

ventral

dorsal

Description and Behavior: The Zebra Swallowtail is a woodland insect found in LA in both mixed coniferous-hardwood forests and deciduous hardwood bottoms. This swallowtail is colonial, although the colonies are not as well defined as other colonial species. Male Zebras patrol about four to six feet off the ground, flying a repeated route, or "orbit" as Lambremont (1954) described it. Zebras seem to prefer flying in more open areas of the forest as well as along roads and trails. When not alarmed, the Zebra Swallowtail has a bobbing flight with shallow wing beats. Once spooked, it usually

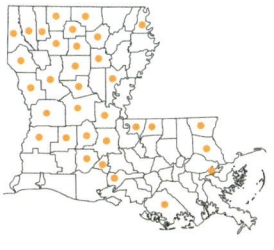

flies away quickly, not to return. It will stop to take nectar, particularly at blooming Brazilian vervain. It is wary and hard to approach, even when taking nectar.

There are at least three broods in LA between Mar and July. Brou (2015b) had records from as early as sunny days in Jan and suggested the existence of multiple broods from Jan all the way into Dec. The first brood is small and lighter colored, with short tails marked by white tips (see dorsal picture). By May, a second, darker brood is flying. This brood has longer tails with the white on the tips extended more than halfway up the tail (see ventral picture). There does not appear to be a clear line of demarcation between the first and second brood, and I have seen lighter-colored specimens with longer tails in late Apr that seem to bridge the gap between the first and second broods. The third brood in July is even darker and larger. Several sources have reported late-summer butterflies into Aug and Sept. For example, one was reported during the Fort Polk count in Sept 2012. These late sightings are rare in my experience.

My Records: I first found this butterfly in the 1990s in the city of Lafayette along the Vermilion River. I also found a large colony at Avery Island, a salt dome in Acadiana that Ross and Lambremont (1963) described as a unique upland-like habitat. I found it over the course of several years in AVO, from late Apr into mid-May, along a slow-moving bayou through hardwood bottomland. In early July, I found several fresh and very dark specimens flying at Chicot SP in a similar habitat. The third brood has been regularly seen at Copenhagen Hills in July (latest date, July 27). In the Weyanoke area, the smaller, spring brood of Zebra Swallowtails is common in Mar and early Apr.

Distribution and Abundance: This is probably the least common resident swallowtail. Its range is primarily restricted by the range of its larval food plant. The plant and the insect seem to be more common along and west of the Mississippi River and primarily north of I-10. It is far less common along the coast. In north LA, it has been found most often during Mar and Apr, at locations such as Walter Jacobs Nature Park and Eddie Jones Park. Other northern locations include

SWALLOWTAILS (Papilionidae)

Zebra Swallowtail (*Eurytides marcellus*) (continued)

Bayou Macon WMA (May and Aug), Bodcau WMA (July and Aug), Driskill Mountain and Roy Quad (Apr and June), and Kisatchie NF in WEB (Apr). In the central LA region, it has been reported as early as March 3 and as late as June 20 in the NAT Kisatchie unit (where I see it every spring but very rarely after early June). Clark reported a sighting (supported by a photo) in LAS in early July of 2017 (this late record is not reflected in the distribution map). In the southwest region, Jeff McMillian provided records, by e-mail, from BEAU into mid-July. In the Cajun Prairie area, it was most common in Mar with fewer seen into June, July, Aug, and even Sept. In Acadiana, it has been reported in the Washington area as part of BugStock (an annual weekend-long entomology, natural science, and music event on private property known as "the Farm") in May, June, and Aug, and once during the Indian Bayou WMA count (early June). There are records from EFE, where it has been described as common and flying from Mar to Sept, most commonly from Mar through June. It has been reported only once, a single butterfly in 1978, over the thirty-eight years of the Pearl River count. In southernmost LA, a sighting was reported in the Grand Bois area in Oct.

Host Plants: Common pawpaw and dwarf pawpaw have been reported as the preferred larval host plants. Pawpaw leaves are believed to contain toxins that are difficult for most herbivores to metabolize; however, the Zebra Swallowtail caterpillar ingests them, with the toxins ending up stored in the adult's wings and body, thereby making it potentially unpalatable to birds.

Black Swallowtail (*Papilio polyxenes*)

ventral

male dorsal

female dorsal

Description and Behavior: This swallowtail prefers open areas like pastures, roadsides, levees, pipeline or power-line cuts, and gardens. It loves flowers and can be easily enticed into your garden by planting one of its several host plants, particularly dill. A favorite nectar source is Brazilian vervain, but it is not picky, taking advantage of whatever is blooming, whether low-growing or tall. When visiting a flower, it hovers, beating its wings rapidly while using its long proboscis to feed. It remains wary while doing so.

This butterfly is dimorphic. The males are smaller with a larger band of yellow spots across the middle of the dorsal hindwings, followed toward the hindwing edge by a small band of metallic blue (sometimes missing), then another band of smaller yellow spots along the outer margin of that wing. In contrast,

Black Swallowtail (*Papilio polyxenes*)　　　　　　*(continued)*

male, form pseudoamericus, dorsal

the female has a small (sometimes missing) band of yellow spots across the middle of that wing, followed by a much larger section of metallic blue, then the same final row of yellow spots along the margin. The female is thought to be a mimic of the Pipevine Swallowtail.

I believe there are two broods and a partial third. The first is in early spring, flying into late Apr. The second starts in late May and goes into June and July. The third flies in Aug into Sept. The third brood is never as numerous as the first two. At Thistlethwaite WMA, the males patrol, flying about five feet above the ground, back and forth along the ditches or in the pipeline cut. The females are equally visible, visiting flowers. When the two meet, they often swirl upward as Buckeyes are prone to do, appearing to be in an aerial dogfight like old World War I biplanes, only to separate and fall back to the ground.

My Records: My records are primarily from west of the Mississippi River. In Acadiana, those records span from early Apr into Sept. The Black Swallowtail is very common, if not abundant at times, at both Thistlethwaite WMA and Indian Bayou WMA in late Apr and May. It is easily found along the levee road at the latter, taking nectar at purple clover growing along the levee. At Thistlethwaite WMA, I find it at flowers in the ditches that line the several roads throughout the WMA, as well as in the pipeline cut 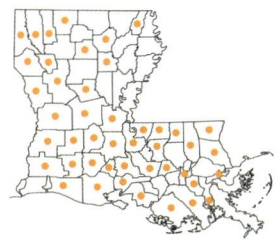 that is such a wonderful location for butterflies and skippers. In central LA, I have seen it from late Mar through late June/early July, with a late male seen in mid-Oct. Over one hundred were counted during the 2017 Catahoula NWR count in mid-May. In southeast LA, I have found it at Abita Creek Flatwoods Preserve.

In early May 2010, at Thistlethwaite WMA, I found an unusually yellow Black Swallowtail, *Papilio polyxenes asterius*, form pseudoamericus Brown 1942, with wide yellow bands equal to those of the western Anise Swallowtail (*P. zelicaon*) and the Western Back Swallowtail (*P. machaon brucei*) (see picture). Philip Wallace sent me a video of a specimen of this form taken by him on May 29, 2004, at Bayou Sauvage NWR. Although this form may occur anywhere throughout the Black Swallowtail's range, it is rare.

Distribution and Abundance: Lambremont (1954) called this swallowtail the "Common American Swallowtail" and noted it was "rather abundant" and "not confined to any particular part of LA." In CAD, dates range from late Feb into Sept, with a single sighting in Nov, most common from Mar to mid-June. Clark reported a sight record for CON in mid-June of 2017 (this late record is not reflected in the distribution map). In the Cajun Prairie region, survey results yielded records from Mar to Sept, with high numbers in Mar and June through Aug. In WFE, records show it on the wing from early May until early Aug with two broods, rare in the blufflands. At Asphodel Plantation, survey records indicate a flight from Feb through Oct, most common in July and Aug. In southernmost LA, there are records from Grand Isle, Galliano, Golden Meadow, and Larose, with multiple sightings in Jan and Feb, then extending into Aug and one sighting in Oct.

Host Plants: Numerous plants have been identified as available in LA, including fennel, parsley, Queen Anne's lace, dill, parsnip, carrots, caraway, and American wild carrot.

SWALLOWTAILS (Papilionidae)

Eastern Tiger Swallowtail (*Papilio glaucus*)

ventral

dorsal female

dorsal female

Description and Behavior: I consider this to be the most common swallowtail in LA, reported from all of the state's regions. It is strongly dimorphic with a black-form female that exhibits shadows of the tiger stripes for which this butterfly is so well known. I've read of areas in the South where 50 percent of the females are reported to be black. I have not found that to be true in LA. Although not uncommon, black females certainly do not constitute one-half of all females. In fact, Israel (1981) noted that black females were rarely encountered during his survey of the Tunica Hills region. Of 196 Eastern Tigers seen, only 27 were the black form (13.78 percent). The yellow females can be distinguished from males by the extensive metallic blue on the dorsal hindwing, absent on the males. Also, in a small number of females, the yellow is a darker, ochre color.

ventral black female

The Eastern Tiger's flight is high and soaring. The best opportunity to observe it is at flowers, as it perches docilely on the bloom, something it does quite readily, with its wings spread wide. Blooming buttonbush is a magnet for this swallowtail. In the heat of midsummer, when little else is blooming, it can be found at stands of Brazilian vervain. Not just a country butterfly, it will visit city gardens, where it is attracted to lantana, Mexican milkweed, and butterfly bush (particularly purple). The males patrol back and forth along a tree line, pipeline/power-line cut, or even a roadside ditch with flowers, typically at least ten feet or higher off the ground. The females are easiest seen at flowers. This swallowtail is most often associated with open areas near woods, but will also fly within the forest's canopy. At Clark Creek Nature Area, both males and females are regularly seen at the damp sand of the creek bottom deep within those wooded hills. The males will visit puddles, wet spots in the road, and mammal scat. They are very attracted to fresh horse manure, and can be found at the Caroline Dormon trailhead at the NAT unit of Kisatchie NF, where the horse trailers are parked.

My Records: I have seen the Eastern Tiger primarily west of the Mississippi River. My records are from Mar until Oct, although I did see one in LAFA on the late date of November 9. It is a regular at Indian Bayou WMA, Avery Island, and in the Weyanoke area, as well as throughout the various Kisatchie

SWALLOWTAILS (Papilionidae)

Eastern Tiger Swallowtail (*Papilio glaucus*) (*continued*)

units. In the fall, when the ironweed is blooming at Thistlethwaite WMA, it is common and easily approached.

Distribution and Abundance: Existing records are from all months except Jan and Dec. In CAD, records are from Feb through Oct. In the Cajun Prairie region, it is on the wing from Mar until Oct. In both of the Felicianas, it has been reported to be common, Feb through Oct, over three broods. It has been recorded almost every year during the Pearl River count with a high of fifty-nine in 2001. One was still flying at Honey Island Swamp WMA in late Sept 2013. In southernmost LA, there are records for Grand Isle, Galliano, Golden Meadow, and Larose from Mar through Sept.

Host Plants: Across its southern range, it uses tulip and yellow poplar, black cherry, ash (Carolina, green, and white), hazel alder, sweet bay, river birch, camphor, Catawba, cucumber magnolia, chokeberry, cottonwood (eastern and swamp), crabapple, hawthorn, hazel nut, hickory, common hop trees, peach, plum, prickly ash, mock orange, spicebush, sassafras, and willow. The caterpillars of this species, the Spicebush Swallowtail, and the Palamedes Swallowtail appear to mimic a snake's head during the later instars.

Spicebush Swallowtail (*Papilio troilus*)

ventral

dorsal male

dorsal
female

SWALLOWTAILS (Papilionidae)

Spicebush Swallowtail (*Papilio troilus*) (continued)

Description and Behavior: This is another swallowtail that primarily inhabits the wooded regions of the state. Its flight is low and direct, as if it has somewhere to go. Males patrol, but not in what would appear to be a recurring route. Females are seen less often, typically either at flowers or circling a sassafras bush. Spicebush Swallowtails are slightly dimorphic. The two sexes can be differentiated in the field, with the males showing a duller green on the lower dorsal wings while the females show a brighter, metallic blue. The female is reported to be a mimic of the Pipevine Swallowtail. Female Pipevines, female Spicebush, and female Black Swallowtails can be difficult to differentiate in the field. On the average, of the three, the female Spicebush will be largest, with its tails resembling baking spoons. The other two have straight tails. Next, look at the spots on the upper dorsal wings. If one row of white/off-white spots is present, it is a Pipevine. If one row of yellow spots is present, think Black Swallowtail. If there are two rows of white/off-white spots, it is a Spicebush.

My Records: My records are mostly west of the Mississippi River, with dates from late Mar into Oct. It is typically found in moister areas where its food plant is also located; however, it is a strong flier and will disperse into more upland, drier habitat. In Acadiana, it is abundant at Avery Island, Indian Creek Rec Area, and Chicot SP. Those particular locations better fit the standard habitat with a lot of water and deciduous woodlands. In the summer, when the buttonbush is blooming, the numbers of this swallowtail can be quite high at Avery Island. In contrast, I see it on an uncommon basis at Thistlethwaite WMA and Indian Bayou WMA, both of which have similar habitat. I do not find it as often in the central LA region. An exception is the GRA unit of Kisatchie NF, where it can be the most common swallowtail in the spring, particularly along the road to Stuart Lake. I have found it at Sam Houston Jones SP in Apr, and it is a regular around the pitcher-plant bogs in VER. It can be extremely abundant at Sicily Island Hills WMA (early June) and Copenhagen Hills (late July). In the southeast, it was present at Sandy Hollow WMA in Mar and Sept.

Distribution and Abundance: Lambremont (1954) believed it was the most common swallowtail in the state. It has been recorded in CAD from mid-Mar to mid-Oct, more common in Apr, June, and July. Other locations from north LA include Driskill Mountain (Apr) and Sparta Quad (July), Lake Claiborne SP (Apr), Kisatchie NF in CLA (Apr), and Bodcau WMA (Mar–Apr and July–Sept).

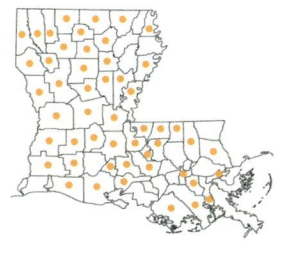

In the Cajun Prairie, records were from Mar through Oct, with highest numbers in Mar, June, and Aug. In southwest LA, it has been reported at Palmetto SP in Apr. There are records for both Felicianas, where it is common, ranging from late Feb into Oct over three broods. Israel (1981) noted that it is attracted to several different flowers (he counted twelve) as well as mud. In south LA, it flies from Feb to Oct. It has been recorded during the great majority of Pearl River counts, with highs of twenty-six (1996), twenty-eight (1999 and 2003), and thirty-one (2001). This swallowtail has also been seen during the Indian Bayou (thirty-four, June) and Fort Polk (Sept) counts.

Host Plants: In LA, this swallowtail is reported to use camphor, sassafras, tulip poplar, spicebush, red bay, and swamp laurel.

SWALLOWTAILS (Papilionidae)

Palamedes Swallowtail (*Papilio palamedes*)

ventral

dorsal

Description and Behavior: I grew up in west TN, and while I saw and collected many kinds of swallowtails, I never saw this one. It was not until I started collecting again in the early 1990s that I encountered it. The first one flew by one day as my son and I were walking along the Vermilion River within the Lafayette city limits. I netted it without knowing what it was. It turned out to be a female, and I was amazed at her large size.

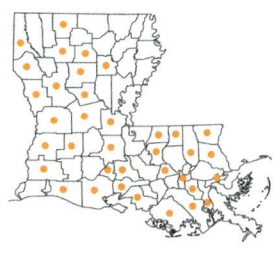

This swallowtail flies low, less than four to five feet above the ground, in a direct fashion, similar to the Spicebush. Once disturbed, it flies swiftly, in a random fashion, often abruptly changing direction, making it hard to follow. It will readily visit flowers and is particularly fond of buttonbush and ironweed, but will also visit low-growing flowers in the spring. The male clearly patrols while looking for a mate. Although it is not a great puddler like the Eastern Tiger Swallowtail, I have found it at damp spots, particularly around old campfires. The males are also attracted to horse manure. The Caroline Dormon trailhead in the NAT unit of Kisatchie NF, where the horse trailers are parked, is always a good spot to look for this butterfly.

My Records: I've encountered the Palamedes Swallowtail primarily in the central and western portions of the state. I have found it as early as March 3 in the NAT unit, and those seen in Mar and into early Apr can be quite small, approximately the size of a midsummer Black Swallowtail. It continues to fly throughout the summer into late Oct. By then, it can be very large, exceeded in size only by Eastern Tiger and Giant swallowtails. While it seems to be present throughout its entire flight season at Bayou Teche NWR and Avery Island, it seems more common in the spring in the NAT, VER, and RAP Kisatchie units. In southwest LA, it was somewhat common once in mid-Apr, at Sam Houston Jones SP. Generally, it flies in areas near water; however, at the three referenced Kisatchie NF units, this swallowtail also flies in the drier, upland areas. I saw a large female in mid-Aug 2017, in the Enon/WAS area.

Distribution and Abundance: Lambremont (1954) believed this was "a rather rare insect in LA," with few south LA records, and it does appear to be rare in north LA. In CAD there is a single sighting in July. Other north LA sightings include Sparta Quad (July) and in WIN (Apr). It increases in occurrence in the central portion of the state, with an early sighting in mid-Feb in NAT. In the Cajun Prairie it was recorded from Mar until Oct, most common in Apr, July,

Palamedes Swallowtail *(Papilio palamedes)*   *(continued)*

and Sept. In southwest LA, there is a colony at Palmetto Island SP, and I saw one in early Oct farther south in the marsh around Freshwater Bayou Lock Rec Area. In EFE, it also appears to be rare, flying in Sept and Oct only. In contrast, this seems to be the most common swallowtail in extreme east LA. For example, Brou (2007d) reported that forty to fifty specimens were captured in ultraviolet-light traps at his Abita Springs study site. There are at least three broods, and there may also be a partial fourth brood. The first three are in the early spring (Apr), late spring to early summer (late May into June), and late summer (late July into Aug). The partial fourth brood appears in late Sept to Oct. It has been regularly reported during the Pearl River count. Across southernmost LA, it becomes less common again with limited sightings reported at Grand Isle and Galliano in Mar, May to July, and Sept. The only records for TER were in early Mar and late Nov.

Host Plants: In LA, the food plants have been listed as red bay and swamp laurel. Brou (2007d) reported rearing it numerous times on swamp bay, a small tree common in the Abita Springs area.

Giant Swallowtail (*Papilio cresphontes*)

ventral

Description and Behavior: The sight of this large, magnificent insect floating on the breeze is always breathtaking. The specimens in early spring are significantly smaller than those seen later in the summer, some as little as half the size. Typically, the males are smaller than the females, and the males have thinner forewings that are somewhat falcated at the tip. In contrast, the females have broader and squarer forewings. By late summer, the females are probably the largest butterfly on the wing, matched only by the largest black female Tiger Swallowtails. Because of their size, they hover above a flower rather than actually land on it. If you approach too closely, they quickly float away but usually not too far. This swallowtail is very fond of buttonbush blooms, and when that bush is blooming around the rookery at Avery Island, many can be present. It will also come to lantana and Mexican butterfly weed, both of which will bring it into your garden, particularly if there is also a citrus tree present. It may occasionally be found at mud and dung.

Giant Swallowtail (*Papilio cresphontes*) (*continued*)

dorsal

I believe there are at least three broods, possible even four. The males patrol, constantly moving back and forth over the area they have staked out. For example, at Thistlethwaite WMA, I have noted the same males patrolling a pipeline cut, occasionally stopping at flowers but then quickly moving on again, only to return a few moments later from the opposite direction. The females are not as secretive as female Spicebush Swallowtails, among others, and are often seen taking nectar at numerous kinds of flowers.

My Records: My records are from throughout Acadiana, central LA, the FL Parishes, and even CAM, with sightings as early as Feb into Oct. Although I have seen it throughout the state, Thistlethwaite WMA remains the most reliable location, particularly in the early fall when the ironweed is blooming. Both males and females can be found in numbers at that flower's tall purple blooms.

Distribution and Abundance: Historically, this swallowtail was reported as more common in south LA with dates from early Mar through Nov. In north LA, it is listed as uncommon to common in CAD from Mar through Oct. In the Cajun Prairie it is present in Mar and Apr, then from June until Oct, most common in June. It is uncommon to common

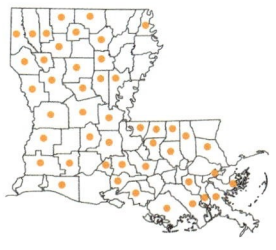

in the Felicianas with sightings from mid-Mar into Nov over three broods. It has been a regular over the years at the Pearl River count. In the River Parishes, it was recorded at Edgard. In southernmost LA, numerous sightings have been recorded from Feb to Nov.

Host Plants: In LA, it has been reported to use American bladder nut, citrus (orange, lime, lemon, tangerine, and grapefruit), common hop trees, Hercules club, and prickly ash trees. I've seen it ovipositing on grapefruit trees and mock orange tree.

WHITES AND SULPHURS (Pieridae)

Checkered White (*Pontia protodice*)

ventral male

dorsal male

dorsal female

Description and Behavior: The sexes of this butterfly are dimorphic. The females are more heavily marked with diamond-shaped spots dorsally along the outer margin and then with yellow scales along the ventral wing veins. In early spring, the males somewhat resemble the females, but by late spring, other than a dark spot of the forewing, dorsally the males are unmarked and white. The flight is low to the ground and usually quick. The Checkered White will

ventral female

stop to take nectar at low-blooming flowers, but most of the time it remains on the move. I find it flies lower to the ground than the Cabbage White, but both appear to move with the same purpose of flight.

My Records: I don't consider this white to be common. I've seen it a handful of times over the years, always as singles except on one occasion. Most of those sightings were in the spring or very early summer. For example, in June 2001, I saw a male at Old River. In Apr of 2010, I found several (both males and females) flying in a field in the NAT unit of Kisatchie NF. Others have reported it in that unit, also in Apr.

Distribution and Abundance: Historically, this white was described as uncommon in LA. In north LA, there have been individual sightings in Apr, May, Sept, and Nov in CAD, including at Eddie Jones Park and C. Bickham Dickson Park. Other recorded locations in that region are the Haynesville butterfly gardens, Bayou Pierre Unit in early Nov, and Jackson-Bienville WMA, Bodcau WMA,

WHITES AND SULPHURS (Pieridae)

Checkered White (*Pontia protodice*) (*continued*)

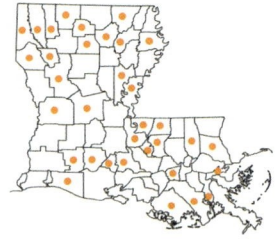

and Cane's Landing in June and Aug. In the central LA region, it has been seen in VER at Vernon Lake in late Oct. Several were reported as part of the Cajun Prairie survey. Those sightings were in May (two seen), July (five), and Nov (two). In southwest LA, this white has been recorded in west CAM in late Apr and then larger numbers in late May. It has been reported several times at Peveto Woods. In southernmost LA, 160 were seen as part of the 2003 Grand Isle count (May). It has also been recorded in LAFO (Apr to Aug and Oct to Nov).

In the southeastern portion of the state, its frequency has fluctuated, described as abundant on one occasion in July in EBR (from three different locations including Greenwood Park during May, July, Aug, Oct, and Nov), but then reported as generally rare in the Felicianas. Records from WFE reflect a flight period from late June to late July and mid-Oct. It has been recorded at Asphodel Plantation in June. Also in the FL Parishes region, there are records from near the Sunshine Bridge in Welcome, near Abita Springs (May) and Fluker. Brou had a few in his collection and communicated that he felt this white was sporadic in its distribution and occurrence in that region.

Host Plants: It was reported ovipositing on peppergrass in CAM in May. In the South, in general, it uses various members of the mustard and peppergrass families, cabbage, cauliflower, and broccoli.

Cabbage White (*Pieris rapae*)

dorsal male

ventral

female

Description and Behavior: Although common elsewhere, this butterfly is in my view not common in LA. I have not seen it here in over twenty years of searching. I cannot explain why the Cabbage White and the Checkered White are not more common within the state. There are no records from the Cajun Prairie or Acadiana regions. It has been found primarily in areas across the eastern portions of the state, sometimes in good numbers, only to later reduce in numbers, occasionally disappearing completely. I am not sure how many broods fly in LA, but clearly there are multiple broods. The spring brood is usually smaller

WHITES AND SULPHURS (Pieridae)

Cabbage White (*Pieris rapae*)   (*continued*)

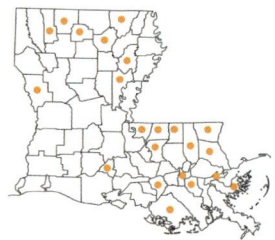

and more yellow ventrally. By midsummer, these butterflies are white. The males have one spot in the middle of the dorsal forewing; females have two.

My Records: The closest to LA I found it was along the Natchez Trace just south of Jackson, MS.

Distribution and Abundance: Lambremont (1954) described this white as "abundant during the spring and early summer months," Jan to Nov. The Louisiana Tech University and University of Louisiana at Monroe collections contain specimens from four northeast LA parishes in Mar, Apr, May, Sept, Oct, and Nov. Loice Kendrick-Lacy, via e-mail, reported it in her vegetable gardens at Haynesville. In central LA, it was once reported as common in SAB, found in a city vegetable garden. In the FL Parishes, it has been recorded on multiple occasions in EBR and WFE (described on occasion as common). In the former parish, locations included Greenwood Park, during Feb, May (multiple sightings), Aug, and Oct. In the latter parish, it was recorded near Weyanoke and Tunica Hills WMA in Feb, Mar, Apr, May, June, and Oct. During his survey of the Tunica Hills region, Israel (1981) had sightings between late Jan and early Aug and then again in late Oct and early Nov. He believed the Cabbage White had three broods in LA, was generally uncommon, but could be "locally common in waste areas near the Mississippi River." It has been reported regularly as part of the New Orleans count (June 1999, 2000, 2002, 2005, 2007, and 2008). Other counts include Pearl (five times, June and July) and Global Wildlife (July). Southeast LA records include Napoleonville in late Jan, at the Amite River in June, in the Edgard area in Apr, and near Holton in Apr. It was recorded in TER in the late 1970s, but it has not been seen there since. Most recently, Patton posted on the LA listserv a picture from the Pine/WAS area, taken in early Aug 2017.

Host Plants: In LA, this white is reported to use nasturtium, ornamental cabbage, and sweet alyssum. In southeast TX, it has been reported to use most wild and cultivated Brassicaceae.

Great Southern White *(Ascia monuste)*

ventral male

dorsal male

dorsal female

Description and Behavior: Other than along the Gulf Coast, this white is generally uncommon in LA, particularly toward the north. Despite that, I've seen more of this white in LA than any other, a reflection of the scarcity of other whites within the state. It is a swift flyer, but will stop often at low-growing flowers. Its flight is typically low to the ground and direct. When taking nectar, it moves about leisurely from flower to flower and is easy to approach; however, once disturbed, it abandons the immediate area quickly. The males are less marked than the females, with the latter often presenting yellow hindwings ventrally with dark scaling along the hindwing veins both dorsally and ventrally.

Great Southern White (*Ascia monuste*) (*continued*)

ventral female

My Records: In Oct 2011, it was very common in CAM with numbers too many to count at Peveto Woods. The 2011–12 winter was very mild, and in June 2012, I saw this butterfly twice within the city limits of Lafayette (I have seen it multiple times in the LAFA area) and, in mid-June, at Kisatchie NF in NAT. By 2013, Great Southern Whites were being reported throughout the state, and during that summer it was the most common butterfly across CAM as well as present at Lacassine Pool. I counted fifty-plus at the Blue Goose Trail, twenty-five-plus at West Cove, and many, many more along the road in Sabine NWR. When I returned to CAM in Sept, I saw several at Brazilian vervain along the Blue Goose Trail, although present in reduced numbers. In Oct 2013, I found three at Freshwater Bayou Lock Rec Area in VRM, and it was still present across south LA during the first two weeks of Nov. In June 2016, I found a large male at the VER-RAP parish line along State Road 463, and within that month, I found males in LAS, IBE, and SMA, all parish records. Others were reported in SAB, TAN, and the New Orleans area during that same summer.

Distribution and Abundance: In north LA, there has been only a single sighting in CAD (Eddie Jones Park, July). In the central LA region, there have been several reports from VER (Allen Acres). In early Aug 2013, a male was photographed at Brazilian vervain in the Cajun Prairie region near Eunice. There are two old reports from the FL Parishes—in

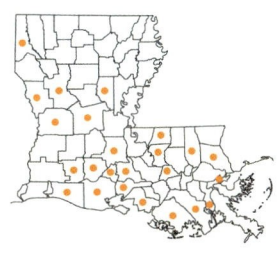

EBR (where it was common in an open field, June) and at Asphodel Plantation (July). Another was reported near Prairieville. In southeast LA, Lambremont (1954) reported once witnessing "many" flying along the shores of Lake Pontchartrain in May. It has been reported regularly as part of the New Orleans count with as many as eight seen in 2005. South of the NO area, it was reported as present by the hundreds in June and July of 1988 around Galliano. Other locations in that region include Golden Meadow, Fourchon, Grand Isle, and Cheniere Caminada, with dates from Mar into Oct. Cunningham recorded two on Nov 18, 2016, in Cocodrie. In Feb and June 1961, it was reported as abundant on Grand Isle in the salt marshes fringing the coast. Other NABA counts that recorded it include Pearl (forty-nine in 1975, twenty-six in 2008), Cameron, Barataria, Fort Polk, and Grand Isle (sixty in 2003).

Host Plants: In LA, this white uses nasturtium, cleome, saltwort, native and cultivated crucifers, and capers. In the South in general, it uses batis along the coast and members of the mustard family inland.

WHITES AND SULPHURS (Pieridae)

Florida White (*Appias drusilla*)

dorsal male

ventral

ventral female

Description and Behavior: This large white is common in south FL and occasionally reported in south TX. Similar in size and markings to the Great Southern White, it is actually more of a woodland butterfly, found primarily in the hammocks along the FL coast into the FL keys.

My Records: My only sighting within LA was in LAFA in May 2001, the day after a severe weather front had moved through the area. I was at an intersection on the outskirts of the city of Lafayette and saw a large white butterfly flopping on the side of the road. I thought it was a Great Southern White, so I put the car in park, jumped out, and picked it up. It turned out to be a female Florida White.

Distribution and Abundance: This white has been reported twice in LA. Other than my sighting, the only other documented record was by Israel in EFE in Oct of 2005.

Host Plants: In south FL, it is reported to use herbs in the mustard family. In southeast TX, it uses saltwort, sea-rocket, peppergrasses, and nasturtium.

WHITES AND SULPHURS (Pieridae)

Falcate Orangetip (*Anthocharis midea*)

ventral male

dorsal female

Description and Behavior: This is the most common white in the northern and central portions of LA, delicately flying from mid- to late Feb into mid-Apr. At Weyanoke, by the second week of Mar, these orangetips are common as they flit in the fringes of thick sections of the forest, staying mostly in the shade and within two to three feet of the ground. Although not necessarily hurried, their flight is direct, with the males searching for females and the females searching for their food plant. Both will occasionally stop to nectar, but my general impression is that they are always on their way to somewhere else. This butterfly is dimorphic. The males show orange wing tips dorsally but not ventrally. The female has no orange whatsoever. I don't believe it can be confused with any other butterfly in LA. Neither the Cabbage White nor the Checkered White is particularly common. Both are larger than this butterfly, show no orange, and prefer to fly in open fields rather than in and along trees.

My Records: On Feb 27, 2011, I found a huge colony (over one hundred) flying in hardwood bottoms of the RAP unit of Kisatchie NF. There had been an ice storm that left a quarter-inch of ice on trees and bushes three weeks earlier; however, during the subsequent two weeks, there had been springlike weather with temperatures of 70 degrees or more. Temperatures during the days before had approached and/or exceeded 80 degrees, including the day I was in the field. I stopped at three spots along Castor Plunge Road, part of which winds for about a mile through a low area crisscrossed with shallow, small bayous. That area was full of blooming wild onions, and the orangetips were at each location, both flying along the road and feeding on the blooms inside the tree line. Both males and females were present in approximately a ten-to-one ratio. This was also the earliest date on which I had seen this orangetip.

In NAT, they can be found flying low to the ground in the hardwood creek bottoms of Kisatchie Hills Wilderness Area. There is also a colony at the end of Forest Road 345 at the trailheads to Kisatchie Bayou. At this spot, they can regularly be found at false garlic. The latest I have seen these orangetips (two males) was April 19, at Copenhagen Hills after a severe winter and late spring. The farthest south I have found this butterfly was at the LA Arboretum in EVA in late Mar and early Apr, in the area of the boardwalk and pavilion on the Wetland trail.

Distribution and Abundance: In CAD, the Falcate Orangetip is abundant from mid-Feb (earliest date, Feb 8) to the end of Mar. Locations in that parish include Eddie Jones Park (fifty-eight were reported in Feb 2013) and Walter Jacobs

Falcate Orangetip (Anthocharis midea) (continued)

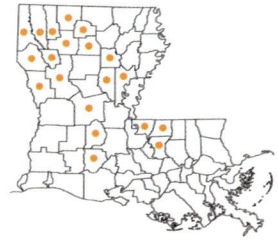

Nature Park. Other north LA locations include Bodcau WMA and Red River NWR (Headquarters Unit) in Mar and Apr, Red River NWR (Yates Tract) in Mar, Driskill Mountain in Apr, and Kisatchie NF in WEB in Mar. A total of three were reported during the survey of the Cajun Prairie area, all in ACA in Apr. Ross (1995) has written an excellent series of articles about discovering this butterfly in the cheniers along the Gulf Coast in CAM. Strickland did not reference it in his manuscript, but his field notes reflect a sighting near Zwolle in early Apr. In the Felicianas, it is reported to be extremely local at Tunica Hills and rare at Asphodel Plantation, from Feb to Apr. It was listed in 2015 as a Tier III, S4-ranked SGCN in LA.

Host Plants: In LA, it has been reported using Pennsylvania bittercress. In the South in general, it uses members of the mustard family including rockcress, bittercress, and peppergrass. In southeast TX, it is reported to use spring cress.

Clouded Sulphur *(Colias philodice)*

ventral

ventral female

dorsal male

Description and Behavior: The Clouded Sulphur and the Orange Sulphur have numerous common characteristics: three dark dots in the postmedian area of the ventral wings; a silver spot in the middle of the lower ventral wing with red rings around it and usually a satellite spot; and a black border on both dorsal wings (solid on males and containing spots on females). The primary differences are color and ultraviolet reflectance. This species is yellow, and the males do not reflect ultraviolet light. The Orange Sulphur is typically orange, and the males do reflect ultraviolet light. Unfortunately, it is not always easy to

Clouded Sulphur (*Colias philodice*)   *(continued)*

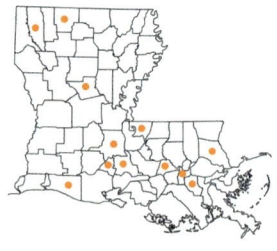

tell these species apart. Particularly in early spring, the Orange Sulphur can be mostly yellow, but will still retain some orange color dorsally in the basal cell region. To further complicate things, the two species are reported to hybridize easily, with these hybrids showing degrees of variance. Also, both of these sulphurs produce white (or alba) females that are extremely difficult to differentiate.

My Records: Most of my records are from the LAFA region in early spring. There was a persistent colony in north LAFA in the 1990s, at Moore Field. On close examination, specimens from this location presented no orange whatsoever. I have seen other early spring specimens elsewhere, including in LAFA, that I concluded were *eurytheme* due to the presence of faint orange at the base of the dorsal wings.

Distribution and Abundance: Lambremont (1954) suggested this butterfly was more of a northern species and rare in its southern range. At that time, he had only one record from LAFA in Nov. Via e-mail, Kendrick-Lacy reported this sulphur as fairly common from spring to fall in and around the butterfly gardens created in conjunction with Haynesville's annual butterfly festival. The photo of this species used in her book was taken just across the state line in AR. The remaining records are from south and southeast LA. In the Tunica Hills region, it appears to be a stray with records from Mar and June. It was reported on the 1993 Bonnet Carré count, as well as during the annual Pearl count, with highs of eighteen in 2005 and thirty-five in 2008. I am unaware of any specimens captured or pictures taken for verification at either of these counts. Brou has specimens in his collection, taken at Edgard (Apr) and in ASC and CAM (June).

Host Plants: In general, in the South, it uses clover and other members of the legume family as its larval host plant. A study by Koehn (2011a) in KY reported that, when presented with different potential host plants, this sulphur "always" used white clover.

Orange Sulphur *(Colias eurytheme)*

ventral

ventral female (alba)

WHITES AND SULPHURS (Pieridae)

Orange Sulphur (*Colias eurytheme*) (*continued*)

dorsal male and female

dorsal female (alba)

dorsal (winter form)

Description and Behavior: This butterfly is usually most common in late spring. After a mild winter, it can be seen as early as Feb. In areas where alfalfa or clover grows, it can be found all year, continuing to breed well into Nov so long as there is no freeze. It flies low to the ground with a direct flight but is prone to stop and take nectar, particularly if there is clover in bloom. It can be very common around fields of clover.

It is dimorphic. The male's dorsal wings have thick, unmarked black borders, while the female's forewing borders contain numerous yellow spots and the borders are much reduced, if not absent, on the hindwings. In early spring and late fall, both males and females can reflect more yellow than orange dorsally, but both will always show a flush of orange at the base of both dorsal wings (but more so on the upper than the lower wing). The alba females are marked similarly to the orange females, but are very white dorsally. Ventrally, they show a light yellow on the hindwings. The Clouded Sulphur is so rare in LA that any confusion between these two should not be a

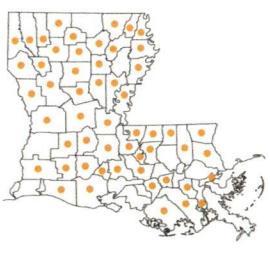

big problem. During the summer, when it is bright orange, the Orange Sulphur might be confused with the Sleepy Orange.

My Records: I've seen it in the LAFA area during every month but Jan. Thistlethwaite WMA and Indian Bayou WMA (along the levee road) are good places to find it during the spring. The alba females can be quite large at Thistlethwaite WMA in May and early June.

Distribution and Abundance: There are records for this sulphur from throughout the state, less common along the coast. In north LA, it has been reported in CAD from every month, common to abundant from Mar through mid-May. Caterpillars have been recorded in Dec in CALD. It has also been recorded in every month in the Cajun Prairie region with good numbers in about half of the months. It is generally common in the Felicianas, with sightings from late Feb into mid-Nov over four broods. Records from below NO include JEF and LAFO during every month but Aug.

Host Plants: In LA, this sulphur is reported to use red clover, partridge pea, and cassia. In KY, Koehn (2011a) raised this sulphur on red clover, white clover, alfalfa, and crown vetch. One conclusion he reached based on his study was that each generation (up to six) produced seasonal forms and that certain forms were associated with larval food plants.

WHITES AND SULPHURS (Pieridae)

Southern Dogface (*Zerene cesonia*)

ventral

ventral (rosa)

dorsal female (top) and male (bottom)

Description and Behavior: I have always thought it odd that this unique sulphur was not more common in LA. It is slightly dimorphic. The male shows bold, black margins on both dorsal wings, while the female usually has yellow spots within the forewing border, and the hindwing border is much reduced or absent. Also, the male's forewing tip is pointed, almost in a falcate fashion. The female's forewing tip is more blocked. The dogface impression in the black border of the upper dorsal wing is less distinct on the female. A rosa form is typically found in the fall and is quite pronounced in some individuals. The

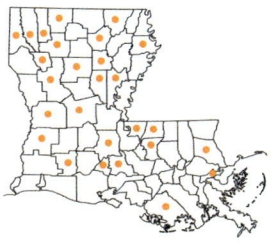

males patrol low, just above the vegetation. The females will settle in the grass to roost. At Copenhagen Hills, numerous examples can be seen in a day as they move about in a fashion similar to the Little Yellows with which they fly. Both sexes readily and regularly visit flowers. When disturbed, both will fly up and away like a Cloudless Sulphur.

My Records: I have typically seen lone singles in the central and southern portions of the state, with these sightings few and far between. Most of my sightings were in the fall, but, in 2012, I saw a male (rosa form) in the RAP Kisatchie unit in mid-Mar. I also found it in the WIN Kisatchie unit in Mar, the NAT Kisatchie unit in Apr, Catahoula NWR in mid-May, and at Sicily Island Hills WMA in June. In mid-June 2013, I visited Copenhagen Hills, where I counted 175 flying in the open glades just as they do at Rick Evans/Grandview Prairie WMA. When I returned in late July, on a gray and overcast day, I counted 77.

Distribution and Abundance: It has been reported as uncommon in CAD (C. Bickham Dickson Park), with occasional sightings between Mar and May and then again between Aug and Nov. Other locations across north LA include Bodcau WMA, Red River NWR, the Bayou Pierre Unit, and Kisatchie NF in RAP, WIN, and WEB. In the central LA region, during the Fort Polk counts in 2012, 13 (Apr) and 22 (Sept), respectively, were reported. McMillian reported regular sightings in BEAU through the fall into early spring, including a late Jan sighting. A July individual was reported from Catahoula NWR at Willow Lake. Only three were recorded during the Cajun Prairie survey, all in JFD (May). Southeastern records include EBR in Oct along the River Road and then extend into the pine flats of the FL Parishes in Aug. Records from the Felicianas (Asphodel Plantation, Ethel, and Jackson) suggest it might only be a stray in Oct and Nov. It was reported on the 2005 (June) New Orleans count, and at Honey Island Swamp WMA in late Sept 2013, only the third sighting there over thirty-eight years. Farther south, there was a mid-Jan sighting in TER.

Host Plants: In the South in general, it uses members of the dalea, clover, and vetch families, false indigo, and alfalfa.

WHITES AND SULPHURS (Pieridae)

White Angled-Sulphur (*Anteos clorinde*)

ventral male

Description and Behavior: This very large sulphur is part of the fauna of the Rio Grande Valley and Mexico. It is a high flier, cruising up and down dirt roads through disturbed wooded areas. Its flight is in a "flip-flop" fashion with its wing position moving from V-shaped to reverse V-shaped, and it seems to bounce along while in flight. It will readily visit large flowers like oleander and hibiscus.

My Records: Within the continental US, I have seen this large butterfly only in south TX.

Distribution and Abundance: The only record for this butterfly was reported by Trahan (2009b) on Sep 30, 2008, in WIN in the calcareous prairies of Kisatchie NF, on Coldwater Creek Road near Coldwater Creek Prairie.

Host Plant: In south TX, it is reported to use partridge pea, a member of the Cassia family.

Cloudless Sulphur *(Phoebis sennae)*

ventral female

dorsal female

ventral male

Description and Behavior: This large sulphur is common during much of the year, flying during even the hottest days of summer. It can be abundant in the fall (Sept and into Oct). During mild winters, it can still be on the wing from Dec to Feb. Brou (2007a) reported the subspecies in LA to be *P. s. eubale*. It is a strong, fast, and high flier and reportedly migratory, although I have never seen reports of any concerted directional movement within the state. Possibly the higher numbers in the fall reflect butterflies migrating into LA from the north.

If you are familiar with the "Peanuts" cartoon strip, and Snoopy's friend Woodstock, then you will understand when I describe this butterfly's flight as "Woodstock-like." As it flies directionally, its actual course is simultaneously all over the place. The best place to observe it is at flowers, which it visits often.

Cloudless Sulphur (*Phoebis sennae*) (continued)

dorsal male

Israel (1981) reported finding it at fourteen different species of flowers. It will also congregate at mud puddles. It is slightly dimorphic. The females are marked with a slight dark brown/black border on the dorsal forewing and there is a distinct brown spot in the middle of that wing. In the fall, the females are often marked ventrally with many brown/orange spots, dashes, and smudges. The alba form, in flight, appears white, but upon close inspection, the coloring is more off-white with a light greenish tint. The males are unmarked dorsally and bright lemon-yellow in color.

My Records: I consider this butterfly to be LA's most common sulphur. I've seen it in every month of the year, throughout the state, no matter the habitat. It prefers open, disturbed areas. While I've never found it in deep woods, it can be seen along the edges of those woods. It can be quite common within city limits and will readily visit gardens and, in particular, red flowers. If I had to name one "best" place, I would go back to my favorite pipeline cut at Thistlethwaite WMA in the fall while the ironweed is blooming. There, it will stay put long enough to approach.

I found a dwarf male in AVO. A specific species's size, as measured by wing length, is an inherited quality, but occasionally dwarf forms are found. These size aberrations can be environmentally induced or genetic, reflecting a recessive gene. I have no idea what caused this dwarf, but the oddly shaped wings suggest it was a result of a problem during the pupae stage.

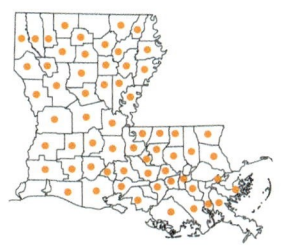

Distribution and Abundance: In CAD it has been reported from every month, most commonly from Mar to early May and then again from late Aug into early Nov. In the central LA and Cajun Prairie regions, it is extremely common, particularly in Aug and Sept. For example, on one occasion, forty specimens were recorded in Aug in the Leesville area. It was the fifth most common butterfly during the survey of the Cajun Prairie region (over five hundred seen), with records from every month of the year. It is abundant in the Felicianas with four broods, flying all year. Surprisingly, this sulphur has been taken at light traps in SJB and STA (with over twenty taken near Abita Springs). In southeastern LA, it is also present during every month.

Host Plants: In LA, this sulphur will use candelabra plant, cassia, and partridge pea. It has been recorded ovipositing on sickle-pod in STA.

WHITES AND SULPHURS (Pieridae)

Large Orange Sulphur (*Phoebis agarithe*)

ventral female

ventral female [alba]

ventral male

dorsal male

Description and Behavior: This brightly colored, large sulphur is a resident of south FL and the Rio Grande Valley which occasionally strays into LA. It is a strong flier and is known to stray well beyond the valley. Although similar in flight to the Cloudless Sulphur, it is distinguishable, even at a distance, due to the large amount of orange presented during

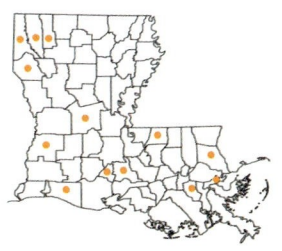

flight. It has been reported from both the eastern and western portions of LA, which causes me to wonder if it might be migrating from both FL and TX.

My Records: I've seen it once within the state, a tattered male found at Indian Bayou WMA in Sept 1995. I also saw a female in Oct, just north of Natchez, MS, on the Natchez Trace, only a couple of miles east of the LA state line.

Distribution and Abundance: There are three old records, all from south LA. More recently, it has been seen several times in north LA. In CAD, it was recorded in Sept of 2007 and 2016 in the Shreveport area. Other locations include Cane's Landing and Cross Lake, in Sept of 2007 and 2016, respectively. A single was reported near Minden in WEB in Sept 2016, and McMillian, via e-mail, reported one in Aug at his nursery in BEAU. Also in southwest LA, one was counted during the 1995 Cameron count (July). Two others were recorded in CAM in Oct, one at Peveto Woods. In central LA, one was recorded during the 2016 Alexandria count (Oct). In southeast LA, it has been recorded three times in Sept—once on the 2002 Folsom count, one at Asphodel Plantation, and one in SCH.

Host Plants: In southeast TX, it is reported to use woody legumes and possibly senna species. In south FL, it uses blackbead and cat's claw.

WHITES AND SULPHURS (Pieridae)

Orange-barred Sulphur (*Phoebis philea*)

ventral male

dorsal male

ventral female

Description and Behavior: This large, colorful sulphur has shown up more often around the state during the last seventeen years. It flies high like Cloudless and Large Orange sulphurs, but with more of a flip-flop flight like the White Angled-Sulphur. This butterfly is dimorphic. To complicate things even further, there are several alba forms. The females appear primarily in varying shades of the albino form, some showing more pink than orange.

My Records: During the fall of 2006, I saw it in LAFA from Sept until Nov at a location with several cassia trees in bloom. I had initially noticed large white sulphurs feeding at the blooms. While I had previously seen this butterfly in south FL and south TX, those seen were primarily the distinctive males and typically colored females. The whitish-colored sulphurs seen in LAFA seemed too large to be Cloudless Sulphurs. I thought they were female Large Orange Sulphurs; however, they ultimately proved

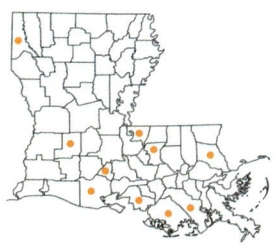

to be this sulphur. Over the course of that fall, I began to see both males and females. They were always in the immediate vicinity of the cassia trees, the males circling and the females feeding at blooms. I witnessed the females ovipositing on the leaves just below the blooms. I saw at least twenty before the first frost in Nov, and none after that frost. I hoped the following spring would bring more, hatched from the eggs laid that fall, but I did not see it further.

Distribution and Abundance: There are three old records from ORL (July and Oct) and ALL (June). Since 2000, the Orange-barred Sulphur has shown up in south LA on a recurring basis. During Oct of 2000, this sulphur appeared in NO in sufficient numbers that Linda Auld reported, via a posting to the LA listserv, that she was able to raise caterpillars, feeding on three types of cassia that she had in her yard—winter cassia, candle bush, and sickle-pod. In Aug of 2004, it again appeared in the NO area and began to breed regularly. Other records from the NO area include fifteen reported as part of the Bonnet Carré count in 2009. It was also reported as part of the New Orleans count that same year. Another sighting occurred there in the fall of 2010. In 2013, one was reported near Morgan City. More recently, a male was photographed in VRM (Abbeville) in mid-Oct 2016, followed by a female at the same location on the first day of Nov. In mid-Dec, a male was photographed at that same location.

South of NO, this species was reported in TER and LAFO between 2006 and 2009, and then again in the Houma area from Sept into late Oct 2013. In the FL Parishes, in 2013 Ross (2013b) reported several specimens in EBR in June and, as the season progressed, it became common by Sept, remaining in the Baton Rouge area into mid-Nov. Other sightings that year included St. Francisville and around Abita Springs (in mid-Oct).

Host Plants: Numerous kinds of cassia have been reported, including winter cassia, candle bush, and sickle-pod. Texas flowery senna has also been reported as used within the state. In south FL, it is reported as using members of the Cassia family such as coffee senna and royal poinciana as its larval food plants.

WHITES AND SULPHURS (Pieridae)

Lyside Sulphur (*Kricogonia lyside*)

Description and Behavior: A very common butterfly in the Rio Grande Valley, this sulphur is famous for occasional irruptions during which large numbers migrate out of south TX and Mexico. About the size of a Checkered White, this sulphur varies in color from yellow to cream to light green, and sometimes a combination thereof. The females are typically larger. The forewing shape is distinctive, similar to but smaller than that of the Southern Dogface.

My Records: The closest I have found it to LA was in the Corpus Christi area.

Distribution and Abundance: This migrant sulphur was first reported in LA at Cane's Landing, in Oct 2006. In May 2012, there was a second sighting at Eddie Jones Park. Finally, in late July 2013, a third was sighted at Peveto Woods by one of the caretakers, with pictures posted on the LA butterfly listserv to confirm the diagnosis.

Host Plant: In TX, its reported food plant is the Lignum Vitae (Guaiacum) family.

WHITES AND SULPHURS (Pieridae)

Barred Yellow (*Eurema daira*)

ventral female (alba)

ventral

dorsal

dorsal female

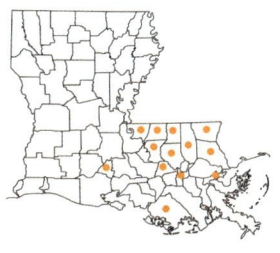

Description and Behavior: This small sulphur flies low, in and just above the grass. It can be difficult to distinguish from the Little Yellow while it is in flight; however, the distinguishing feature is a black bar along the bottom of the forewing, for which this sulphur is named. Unfortunately, it can still be hard to differentiate from the Little Yellow when at rest since both land with their wings held over their backs, making it difficult to see the black bar. This butterfly has two seasonal forms, wet and dry. The wet or winter form has less black and is ventrally brown or red, while the dry or summer form is gray/white.

My Records: I have found it at Sandy Hollow WMA in Mar, Aug, and Sept. It was extremely abundant in Aug and Sept (with well over fifty seen in less than one hour's time).

Distribution and Abundance: The vast majority of LA records are from the FL Parishes, where it can be common in open pine woods. I suspect these records are an extension of those populations along the AL and MS Gulf Coast that have moved west under appropriate conditions. One old record was of a dense population near Covington in early Sept. The LSU State Collection contains some specimens taken in TAN during Sept. More recent FL Parishes records include multiple sightings from Feb to Nov at places such as Woodland, Rogillioville, and Sandy Hollow WMA, as well as Edgard in SJB. It was reported once (June 2000) as part of the New Orleans count, and twice on the Global Wildlife count (July 1994 and July 1995) and Pearl counts (1996 and 2012). South of the FL Parishes, this sulphur was reported from TER in the late 1970s but has not been seen since.

There are old records from the western portion of the state; however, I have questions about all of them. One is a single record from "Lafayette," taken in Sept 1936. As indicated, I would expect any sightings within the state to have been farther east, not in Acadiana, and I wonder if this record might be a mistake. The only information given by Lambremont (1954) was the location (Lafayette), the date, and a parenthetical notation, "SLI," which was defined to mean that particular specimen was in a collection at Southwestern Louisiana Institute (now the University of Louisiana at Lafayette). Because this record and two others from Lafayette, for the Julia Longwing and Great Spangled Fritillary, were for butterflies well beyond their anticipated range, I wonder if the

WHITES AND SULPHURS (Pieridae)

Barred Yellow (*Eurema daira*)   *(continued)*

reference to Lafayette is the collection's location, not where the specimens were caught. Without access to the collection, it is impossible to verify these unusual records.

This sulphur was also reported in the Cajun Prairie survey during the months of Oct to Dec. That survey did not list any Dainty Sulphurs although that butterfly is regularly present in the western part of the state. No pictures or specimens were taken during the survey, and the participants have subsequently agreed those sightings were probably Dainty Sulphurs. Finally, one Barred Yellow was reported on the 2003 Alexandria count. Dainty Sulphurs were also reported on that count in 1999, 2007, and 2010 (three seen). The compiler for that count has indicated the 2003 specimen was not collected, pictures were not taken, and the probability of misidentification was high.

Host Plants: In FL, it is reported to use herbs in the bean family like vetch, pencilflower, and hairy indigo.

Mexican Yellow *(Eurema mexicana)*

ventral (summer)

ventral (fall)

dorsal male

WHITES AND SULPHURS (Pieridae)

Mexican Yellow (*Eurema mexicana*) (*continued*)

Description and Behavior: This western butterfly is prone to long-distance migrations. It is a rare mid-summer and fall stray into the Houston area. Also, it is listed as a member of the butterfly fauna at Red Slough WMA in southeast OK, not far from CAD. It flies above the grass in open, disturbed areas like overgrown fields. It is twice the size of the Little Sulphur, and bright yellow ventrally, so it should not be confused with the Sleepy Orange or the Orange Sulphur. Dorsally, it is cream colored. In LA, it might be confused with the Clouded Sulphur, but its square forewing and dorsal forewing markings in the shape of a small poodle's head outlined in black are diagnostic.

My Records: The closest I have seen it to LA was at Rick Evans/Grandview Prairie WMA in southwest AR.

Distribution and Abundance: There are two old reports of several sightings in Oct and Nov, in CAD. More recently, in early Dec 2007, one was seen at Caddo Lake Dam, and, in mid-Nov 2013, another was sighted at the Shreveport Yacht Club.

Host Plants: In southeast TX, it is reported to use acacia, senna, and members of the Diphysa family.

Little Yellow (*Pyrisitia lisa*)

ventral

dorsal male

dorsal female (alba)

Description and Behavior: This small sulphur is common to abundant throughout the state. It flies low to the ground and readily visits many low-growing flowers as well as mud. Israel (1981) reported seeing it at eighteen different species of flowers. The males are typically smaller, and a dorsal black border

WHITES AND SULPHURS (Pieridae)

Little Yellow (*Pyrisitia lisa*) (*continued*)

ventral female (alba)

extends strongly down the forewing well into the hindwing. That black border is typically reduced on the hindwing of the female. Both sexes have a black spot at the base of the ventral hindwing and a brown square spot on the upper margin of the same wing. The female commonly has more black or gray spots and splotches on the ventral hindwing. The alba-form female can be very common at times, particularly in the heat of late summer when they can be quite large. The Little Yellow is smaller and more yellow than the Sleepy Orange. It is larger and shows more extensive black borders dorsally than the Dainty Sulphur where they overlap in west LA. In the FL Parishes, it can be confused with the Barred Sulphur; however, the latter has a black bar along the bottom of the dorsal forewing (see pictures) that can be seen in flight.

My Records: I don't see this sulphur as often as I do the Cloudless Sulphur, but in terms of total numbers counted each year, I see more of this one. I find it is most common in the fall. While I have seen it in every month of the year, I believe it repopulates LA each year from south TX, typically reaching here in higher numbers during late summer. It clearly breeds in LA, with successive broods flying deep into Nov as long as the warm weather holds. Both Thistlethwaite WMA and Indian Bayou WMA are good places to look for the alba females. I also saw several at Copenhagen Hills in July.

Distribution and Abundance: Lambremont considered this sulphur to be one of the most common butterflies in LA, recorded in almost every parish. In CAD, it has been described as common to abundant from May into Nov. In the Cajun Prairie region, records are from Mar until Dec, but in high numbers only during July, Aug, and then into Oct. It has been reported as extremely abundant in WFE while only common at Asphodel Plantation, over four broods with records in Jan and then Mar to Dec. Records from the FL Parishes include Honey Island Swamp WMA in the fall. In the River Parishes, five were found in ultraviolet light traps in SJB. In south LA, it has been found in every month except July, with an early date of January 10 and a late date of December 31.

Host Plants: In LA, this sulphur uses red clover, vetch, cassia trees, and partridge pea. In the South, in general, it uses members of the senna, pea, and pea vetch families.

WHITES AND SULPHURS (Pieridae)

Sleepy Orange (*Abaeis nicippe*)

ventral

dorsal

ventral [rosa]

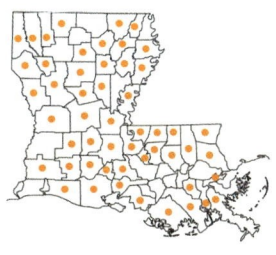

Description and Behavior: This is another sulphur that can be seen during every month of the year but is more common in the fall, at which time it is often the second most common sulphur behind the Little Yellow. It visits many different low-growing flowers—fourteen species counted by Israel (1981)—and participates in mud-puddle parties. It typically flies low to the ground, and the only other sulphur with which it might be confused is the Orange Sulphur. The best way to distinguish the two is to inspect the ventral wings. The Orange Sulphur has a large ringed silver spot in the center of the lower ventral wing with three black dots between that spot and the wing margin. The Sleepy Orange does not have either the ringed spot or the black dots. This sulphur has a spring/summer form that is bright yellow ventrally and a fall, or rosa, form that is pinkish red (see pictures).

My Records: I've seen it in every month of the year, primarily west of the Mississippi River. East of that river, I have seen it in SHE and STA, in addition to both of the Felicianas.

Distribution and Abundance: Lambremont (1954) described this sulphur as "common during most of the summer months." In CAD, it has been recorded during every month except Feb. In the Cajun Prairie region, it was reported in Jan and then Apr to Nov, most common in Aug and Sept with over one hundred seen during each of those months. There are records in both Felicianas, where it is common, flying over three broods from Jan to Dec. In south LA, it has been recorded in the fall into early Dec.

Host Plants: In LA, this sulphur is reported to use red clover, vetch, cassia (such as coffee senna, winter cassia, and sickle-pod), and partridge pea.

WHITES AND SULPHURS (Pieridae)

Dainty Sulphur (*Nathalis iole*)

ventral (summer form)

dorsal

ventral (fall form)

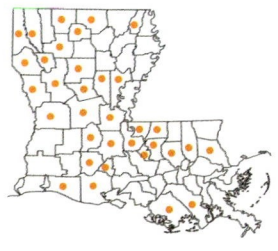

Description and Behavior: This is the smallest sulphur found in the state. It flies low to the ground, just above grass level like the Little Yellow, but can be easily distinguished from the Little Yellow by its smaller size. In early spring and late fall, its ventral hindwings are greenish colored, while in the summer those wings are brighter yellow. While both the Barred Yellow and this sulphur have a black bar along the bottom of the dorsal forewing, this sulphur is smaller and has two black spots on the ventral forewing.

My Records: Other than in CAM, I usually see it as singles. I have found it at Kisatchie NF in NAT in both Apr and May, in good numbers. I also found a single male at Duralde Prairie in late Apr one year, but typically it is not seen until later in the year. I have seen only two in LAFA, in Aug and Sept. I found two, feeding on what I believe to be gumweed, at the Louisiana Tech University soccer complex in LIN in early Aug. In Oct, I have found it at a rest stop along I-10 in SLA and at Freshwater Bayou Lock Rec Area.

Distribution and Abundance: Recorded throughout the state, it is more common in west LA. In CAD, there have been individual sightings in Jan and Feb, occasional sightings during May through July, and then it becomes common from Sept into Nov. Other north LA locations include Eddie Jones Park, the Bayou Pierre Unit (Oct), along Loop Road in UNI (June), Bodcau WMA, Red River NWR, Loggy Bayou WMA, and Cane's Landing (May and July through Nov). In the central LA region, it has been reported in eight parishes, mostly in the fall. The Barred Yellow records from the Cajun Prairie region for Oct through Dec were probably actually this species. There are a few records from southeast LA, virtually all after Aug. It is reported to be uncommon in the Felicianas, flying from early Aug to early Dec, over two broods in sandy waste habitats. In the FL Parishes, it was recorded at the Global Wildlife Center. Old records reported this sulphur as common in fields in the NO area.

Host Plants: In southwest TX, it is reported to use dogweed, sneezeweed, shepherd's needle, and other small Asteraceae.

GOSSAMER-WINGED BUTTERFLIES (Lycaenidae)

Harvester (*Feniseca tarquinius*)

ventral

dorsal

Description and Behavior: The Harvester is uncommon but appears to reside throughout the state in deciduous habitat near water. It can be found perching on the sun-splashed leaves of lower tree branches as well as flying low to the ground or perching on moist ground, including wet roads and streambeds. The

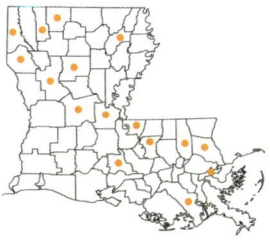

flight is mothlike, almost frenetic. At rest, the wings are held over its back like a hairstreak. Adults are not known to feed at flowers; rather, the adults are reported to feed on aphid honeydew, dung, and sap, and to sip from mud. In FL, a flight period from Feb to Dec has been documented. There are three to six generations in the southern US.

My Records: I found two during the 2009 Shreveport count (June) and a third at Eddie Jones Park (Apr). I have since found individuals in mid-Mar, mid-Apr, and May at the NAT Kisatchie unit in a low-lying area of deciduous woods where I have also found Falcate Orangetips and Southern Pearly-eyes. Also in central LA, I saw two in mid-Mar, in the RAP Kisatchie unit. The farthest south I have found the Harvester was at Indian Bayou WMA, in July, when I was surprised to see two, in different areas of the WMA. Both were in dense deciduous woods near drying creek beds that still had some moist mud. One was actually down on the mud while the other was perched in a tree above the mud. In the southeast, I saw one in Weyanoke, at sand in a moist creek bed.

Distribution and Abundance: This unique butterfly was first reported along the banks of the Ouachita River in dense woods. Lambremont (1954) described it as flying about four feet above the ground, "in a very nervous manner and was taken with difficulty." In CAD, it has been reported as occasional, during May and June (C. Bickham Dickson Park). Kendrick-Lacy reported, via e-mail, two at Miller Lake in CLA. I am aware of specimens from EBR and STA in the southeast portion of the state. It has been reported on one New Orleans count in June 1996. Other southern records include Galliano (June) and in TAN (Mar).

Host Plants: The Harvester is the only carnivorous butterfly in North America. In the FL Panhandle, the caterpillars have been recorded feeding on woolly ash aphids and beech blight aphids. Other reports from FL include woolly maple aphids and woolly alder aphids as common prey. In LA, an association has been suggested between this butterfly and hazel alder and beech trees. The adults spend most of their time in the vicinity of aphid colonies, with the males reported to patrol in and around the host trees. The female lays her eggs on leaves or stems close to those colonies.

GOSSAMER-WINGED BUTTERFLIES (Lycaenidae)

Great Purple Hairstreak (*Atlides halesus*)

ventral

ventral

Description and Behavior: As LA's largest hairstreak, this incredibly beautiful butterfly is actually blue, not purple. The males are usually smaller, show more metallic color, and have smaller black borders on the dorsal forewing. The females will show much more black on the wing borders, as much as one-half of

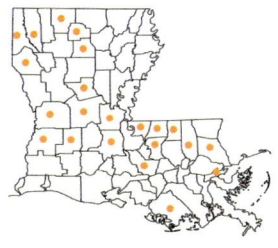

the dorsal forewing. It also appears to me that the males are a brighter blue dorsally while the females reflect a greenish blue.

My Records: My records reflect that this hairstreak is multi-brooded, flying from late Feb into Apr, then again in late May into June, and finally in Sept. I believe there are three broods. It is very rarely seen in numbers, and often only as a single. My first experience with it was in southern AVO (the same location where I first found Banded Hairstreaks). I caught a fresh male on the ground at the base of a large tree with mistletoe in its canopy. Returning to the spot the next year, I found two on the ground at the base of that same tree, all fresh and probably just emerged from the pupae stage. Unfortunately, a severe storm knocked the top out of that tree, killing all of the mistletoe.

I have also seen it on multiple occasions at Thistlethwaite WMA, in the spring and then primarily in the fall. In the spring, it was found taking nectar at blooming blackberry bushes in sunny patches along four-wheeler trails. In contrast, in the fall, it was found out in the open, taking nectar from blooming boneset. I have also seen it at the Catahoula Butterfly Garden in Mar, taking nectar at a blooming pear tree, and in the Weyanoke area, again in Mar.

Distribution and Abundance: Records from around the state include individual sightings from CAD in Mar, July, Aug, Sept, Oct (at Eddie Jones Park), and Nov. It has also been reported in north LA at Cane's Landing, Bodcau WMA (Oct), and the Stonewall area (Sept). I am aware of only two records from the Cajun Prairie (in Sept and Oct), and McMillian, via e-mail, reported one in Oct in BEAU. It is reported to be fairly common in EBR (including Waddill Wildlife Refuge), primarily in Sept and Oct.

WFE records suggest two broods from late Mar to early May, July and Sept into early Nov, rare to uncommon. At Asphodel Plantation, records are from Jan and then from May to Dec, most commonly in Oct. Also in the FL Parishes, it has been recorded at the Global Wildlife Center, and Brou has some large, beautiful specimens from his property near Abita Springs and from the Greensburg area. There is an old record from ORL in Nov, and it has shown up once in 2007 as part of the New Orleans count. In southernmost LA, there is a record from TER in Nov.

Host Plant: In LA, it is reported to feed on semiparasitic mistletoe, usually in elm, hackberry, or live oak trees.

Juniper Hairstreak (*Callophrys gryneus*)

ventral

ventral

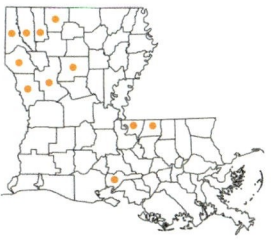

Description and Behavior: This is another of LA's more colorful hairstreaks, with bright green scaling on the ventral wings. It is LA's only documented green hairstreak, so it should not be confused with any other hairstreak. It is most common in the spring (Mar and into mid-Apr).

I'm sure everyone has read or heard of the old trick of shaking a red cedar tree to make this hairstreak fly out, and truth be told, it actually works. The problem is that they usually either fly off or return higher up in the same tree. So, as an alternative, check the area for small white flowers such as clover, wild onion, or blackberry, particularly in the morning or late afternoon. They are easier to approach while at flowers.

My Records: In LA, I've seen this hairstreak at Kieffer Prairie (WIN) in Mar and Eddie Jones Park (CAD) in Apr.

Distribution and Abundance: At Eddie Jones Park, near stands of red cedar, this butterfly can be numerous with records from late Mar all the way into early Sept. Other records from north LA include Kisatchie NF in WEB (Mar and Apr) and Bodcau WMA (Apr, June, July, and Aug). Judging from these records, there might be as many as three broods in north LA, ranging from late Mar into Apr, late May into June, and then a possible third starting in late July through early Sept. These two summer broods are smaller and irregular. In the central LA region, it has been reported in NAT (Apr) as established but local near red cedar stands. It has also been found in SAB in Mar. Bette Kauffman provided a photo of this species, taken in early June of 2017 at Allen Acres in VER (this late record is not reflected in the distribution map). During a chance meeting at Asphodel Plantation, Israel personally communicated to me a flight period in EFE and WFE for this hairstreak of Mar 12 to Apr 14. In that region of the state, there appeared to only be one brood, during which this hairstreak is uncommon, seen at flowers in fields adjacent to stands of its food plant. Israel also related to me that he had once discovered a colony of Juniper Hairstreaks at Avery Island.

Host Plants: In LA, it is reported to use any native juniper, but primarily Eastern red cedar.

GOSSAMER-WINGED BUTTERFLIES (Lycaenidae)

Frosted Elfin *(Callophrys irus)*

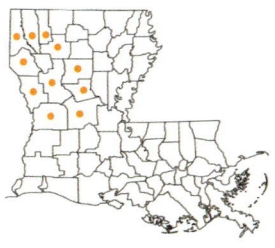

Description and Behavior: The Frosted Elfin has a range that extends across much of the eastern US. Despite this extended range, *irus* is usually extremely local. Living in small, scattered colonies, this butterfly has a single spring flight period in the South, Mar to Apr. There are three subspecies, one of which, *C. i. hadra,* is found in east TX, west AR, and into northwest LA. Several authors suggest *i. hadra* should be recognized as a separate species because it feeds only on wild indigos.

Frosted Elfins were listed in 2015 as a Tier III, S2/S3-ranked SGCN in LA. Its preferred habitat in LA is pine barrens, where its preferred larval host plant, yellow indigo, grows. It rarely strays far from its food plant. At the NAT unit of Kisatchie NF, I have seen Frosted Elfins use false garlic and blooming redbud as a nectar source. When found, it is typically common to abundant. In central LA, this elfin takes to the wing before most of the grass around its food plant has started turning green. Its dark brown color is easy to follow as it flies around and perches on its food plant. Both sexes are low flying and low perching. Its flight is not as fast or frenetic as other hairstreaks, but more bouncy like a small satyr.

Because this elfin is on the wing at the same time as Henry's Elfin and their ranges coincide in LA, confusion may occur. Aside from the subtle differences reflected by the photos, I offer two main distinctions between the two butterflies: (1) *henrici* flies higher and around its food plant, the redbud tree, while *irus* flies low around wild indigo, and (2) *irus* has a small thecla dark spot near the hindwing margin that *henrici* lacks. Occasionally, that spot is hard to discern due to wear. In a few it appears to be completely lacking.

My Records: The flight period ranges from late Feb into mid-Apr. Late-spring cold weather strongly affects this elfin's emergence dates. Typically, they seem to make an initial appearance once the host plant, wild indigo, has grown to between six to twelve inches. The more southern colonies appear earlier than those in CAD. I saw five adults at Eddie Jones Park in mid-Apr, associated with wild indigo. I found a colony in the large open area adjacent to the Catahoula Butterfly Garden and another near the Stuart Lake Recreation Complex. There are several colonies in Kisatchie NF in NAT. The earliest I've seen these colonies is March 3 (in 2012). Finally, I have found colonies at Indian Creek Rec Area and in the RAP (along Castor Plunge Road) and WIN units of Kisatchie NF.

GOSSAMER-WINGED BUTTERFLIES (Lycaenidae)

Frosted Elfin (*Callophrys irus*)   *(continued)*

Distribution and Abundance: This elfin was first listed as part of LA's butterfly community based on an old report by Skinner in 1907, although Skinner identified no location (Lambremont 1954). The next records were primarily from the central LA region. Several specimens were reported during the late 1960s, in and around the GRA unit of Kisatchie NF. During that same time frame, twenty-seven specimens were collected in northwest LA, where it was reported as common in the longleaf pine hills of NAT and DES. Specific locations included Vowells Mill (seventeen seen in early Apr) and near Hunter (five seen, early Apr). In north LA, it has been reported at Walter Jacobs Nature Park, uncommon and flying from as early as Feb 26 to Apr 11. Other records include Bodcau WMA (late Mar) and Kisatchie NF in WEB (Apr).

Host Plant: Its preferred larval host plant is yellow indigo, an herbaceous perennial plant in the family Fabaceae. Yellow indigo prefers dry meadow and open woodland environments, in sun to bright shade. Since it fixes nitrogen, it can tolerate poor soil. The multiple bushy stems of wild indigo reach two to three feet tall. The leaves are silver-green. The flowers are yellow and grow in spikes 1.5 to 3.0 inches long.

Henry's Elfin (*Callophrys henrici*)

ventral

ventral

GOSSAMER-WINGED BUTTERFLIES (Lycaenidae)

Henry's Elfin (*Callophrys henrici*)

(*continued*)

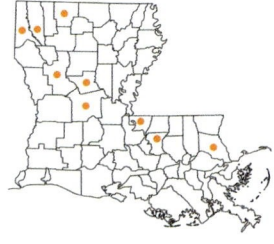

Description and Behavior: The most successful method for locating this elfin is to search for blooming redbud trees along roads and trails during Mar. These elfins don't move around much while at redbud flowers. So, if the tree is tall, I gently tap different sections of the tree with my pole, thereby causing any butterflies at blooms in the tree to fly. If the tapping is gentle, they will move only a short distance before resettling. Their dark color against the pinkish red blooms should allow you to then follow their movement.

My Records: I found that a good place to look for Henry's Elfins is along the top of the ridge where the main trail begins at Longleaf Vista Rec Area in NAT when the several redbud trees there are in bloom. Another good area for Henry's Elfins is at blooming redbud trees in the lower, wetter parts of the Red Dirt Area of the NAT unit of Kisatchie. In the Weyanoke area, I found it from mid-Mar through the first week of Apr. I saw fourteen in that area in early Mar one year, several of which were at eye to ground level. One was at a sunny spot of dirt; three (all males) were perched on underbrush at about eye level alongside a trail through a densely wooded area. Two others were perched directly below a blooming redbud tree.

Distribution and Abundance: Current records reflect a limited range in LA. Henry's Elfin appears to be uncommon in CAD (Soda Lake WMA) during mid-Mar. It was also found in north LA near Homer in CLA (late Feb) and at Bodcau WMA in mid-Mar. I suspect it is present across north LA. Central LA records are from VER, NAT, GRA, and RAP. In the FL Parishes, it has been recorded in EBR (Mar) and Abita Springs, where numerous specimens have been reported.

Host Plants: In the South, it is reported to use redbud primarily, but also huckleberry, farkleberry, American holly, yaupon, dahoon, American plum, and mountain laurel. Other suggested larval food plants include highbush blueberry, black cherry, swamp doghopple, holly, huckleberry, persimmon, plum (American, Mexican), leatherwood, and viburnum.

Eastern Pine Elfin (*Callophrys niphon*)

ventral

ventral

GOSSAMER-WINGED BUTTERFLIES (Lycaenidae)

Eastern Pine Elfin (*Callophrys niphon*) (*continued*)

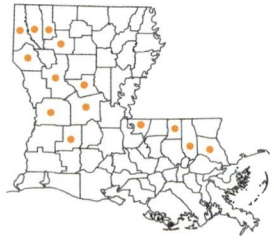

Description and Behavior: Easily recognized by the distinctive banded pattern on its ventral hindwing, this beautiful butterfly starts flying about one week after Henry's Elfin appears. I consider this elfin more common than Henry's because the latter is so much more restricted by the limited distribution of its host plant. Eastern Pine Elfins can be found "puddling" at muddy and wet spots and at old campfires, but you should approach slowly as this butterfly is usually wary. I look for it at about head to waist level, perched on the outer limits of new leaves growing from young longleaf pine trees. On two occasions, I've witnessed one literally fall out of the sky and land at my feet. Both times were during very windy days, and I suspect they had been blown out of nearby pine trees. Adults feed on spring cress and wild plum.

My Records: The Kisatchie NF unit in NAT is a good place to find it, particularly at Little Bayou Pierre, a clear-running stream in the Kisatchie Hills Wilderness Area. I've also found this elfin in Alexander SF and along Castor Plunge Road in the RAP unit. Another good place is the Catahoula Butterfly Garden. There are several types of bushes and trees that bloom with groups of small white flowers in mid- to late Mar, all of which attract this elfin. Most of the time, I see this elfin as singles, but I once saw five on a single bush in that garden. The only two occasions on which I found this elfin outside of the central LA region were in late Mar, a single at Weyanoke and two females at Sandy Hollow WMA in STA.

Distribution and Abundance: This elfin's range and flight period within the state are restricted. In CAD, it has been reported from Mar into early Apr at Eddie Jones Park. Other locations in north LA include Kisatchie NF in WEB (Apr), Jackson-Bienville WMA (Mar), and Bodcau WMA (Mar). Other LA records include ALL (Mar) in the southwest and VER (Apr) in central LA. In the FL Parishes, it has been reported near Greensburg and near Abita Springs, from the first day of Mar through the end of that month.

Host Plants: This elfin uses members of the longleaf pine family, generically listed as "pine" in southeast TX. It presumably uses those same trees here in LA as I have always found it flying in and around longleaf pines.

Oak Hairstreak *(Satyrium favonius)*

ventral [No. ssp]

ventral [No. ssp]

GOSSAMER-WINGED BUTTERFLIES (Lycaenidae)

Oak Hairstreak (*Satyrium favonius*)   (*continued*)

ventral (So. ssp)

Description and Behavior: In 1951, Klots described this hairstreak as two separate species, the Southern Hairstreak, *Strymon favonius*, and the Northern Hairstreak, *S. ontario*. Eventually, they combined into one species, the Southern Oak Hairstreak, *Fixsenia favonius*, with several subspecies, including *ontario* in the East and *favonius* in the Southeast. It has now been moved into the Satyrium family and given the common name "Oak."

It is single-brooded, flying in late Apr and May. I have found it most often during the last ten days of Apr; however, Brou (2007a) reported a peak flight time toward the end of the first week of May with continued activity into June. The Southern subspecies can be easily confused ventrally with White M Hairstreaks. (The Northern subspecies is much smaller than the White M.) Of course, the blue dorsal side of the White M is diagnostic, but when the hairstreak is perched and only showing its ventral sides, the two are quite similar. If you can get a close look, the extension of the red spot (actually a streak of red on the Southern) up the length of the sub-margin of the hindwing will differentiate the two. The Southern subspecies is also similar to the Gray Hairstreak, but I find the background color of the Southern to be more of a brownish gray, and the Gray lacks the blue scaling between the red anal spots. The flight of both the Southern and the Northern subspecies is typical of hairstreaks, fast and erratic, but they are quite docile when at flowers.

My Records: I've found both the Northern and Southern subspecies in LA. In both GRA and WFE, I have found only the Northern. At Avery Island, I've found only the Southern. In between, at Thistlethwaite WMA, I've found both on the same day and on the same tree. The colony at Avery Island numbers in the hundreds in late Apr and early May.

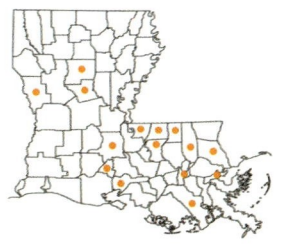

They are primarily found at a long row of flowering ligustrum along the bayou. This colony has the most extensive patches of red extending up the ventral hindwing as I have ever seen. Conversely, the specimens in GRA are much smaller and with minimal red on the hindwing. At Thistlethwaite WMA, the individuals that I believe are the Southern subspecies are smaller than those found at Avery Island, with smaller dorsal orange patches. At a ratio of about sixty to forty in favor of the Southern, the Northern specimens have no orange dorsally and minimal red ventrally.

Distribution and Abundance: This hairstreak was initially reported in LA in 1954 as *S. ontario* based on one record from ORL in May. Most of the older records are from the FL Parishes. In the 1960s, additional records, identified as *Eurystrymon ontario*, were reported from EBR, including Greenwood Park, where several were witnessed in May taking nectar in the upper branches of a privet hedge. There are two records from the central LA region, at Kisatchie NF in NAT (May) and near Many in SAB (mid-June), and R. Patterson reported in the 2016 Lepidopterists' Society's Season Summary a sight record for Kisatchie NF in NAT in early May (the Patterson report was received late and is not reflected in the distribution map). It has been recorded as rare at Asphodel Plantation in Mar–Apr and June–Nov. South of NO, *ontario* was reported from early to mid-Apr in 1990 in the Galliano area.

Host Plant: Oak is reported to be the food plant.

GOSSAMER-WINGED BUTTERFLIES (Lycaenidae)

Banded Hairstreak (*Satyrium calanus*)

ventral

ventral

Description and Behavior: This hairstreak is single-brooded, flying from late Apr to early June,
primarily during the first three weeks of May. Brou
(2009a) reported *calanus*'s flight period to peak
about three weeks earlier in STA than the "similar
looking *S. kingi*." My records reflect that Banded
Hairstreaks seem to fly about two to three weeks

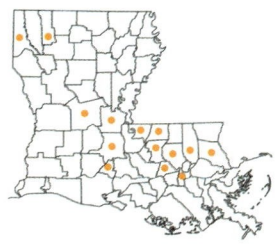

before King's, although on one occasion there was overlap. All of the specimens
I've seen in LA were the darker variant. Typical of hairstreaks, flight is swift and
often difficult to follow. The males will continue to fly late into the afternoon
(after 5:00 p.m.).

My Records: My first experience with this hairstreak was at a location near
Chicot SP. I was walking along a deer hunter's trail and came to a crossing with
a similar trail. At the junction, the sun was shining through the trees and several
males were engaged in aerial dogfights as they jostled for the choice spots at
the end of lower branches that extended into the sun. I later witnessed these
same "dogfights" at Indian Creek Rec Area in early May in a section of deep,
hardwood swamp not easily accessible without boots and protective clothing. I
found this spot by chance by following a hunter's trail marked by orange tape.
(I love following hunter's trails; they lead through the darnedest places.) I first
noticed dogfights when I saw flashes of silver and blue as the males whirled
through patches of dappled sunlight in the understory. Between battles, they
landed on the lower extended branches of elm trees. I returned in late May of
that same year to find numerous King's Hairstreaks in the same area along with
a female Banded.

I see this hairstreak almost yearly at Thistlethwaite WMA. Typically, they
have been found at Chinese privet bushes that bloom in Apr. In 2006, I found
numerous males perched on the tips of tall grass growing around one of the
oil-production pads found at Thistlethwaite. This odd perching behavior by the
males went on for several days. It was very windy on each of those days, so I
suspect this behavior was related to the wind. I've not witnessed this behavior
since. I have also seen it at Bluebonnet Swamp Nature Center (in EBR) in Apr
and at Mary Ann Brown Preserve (in WFE) in late May.

Distribution and Abundance: The first record for the Banded Hairstreak in
LA was from EBR in Apr and May (including Greenwood Park in early May).
In north LA, there are occasional CAD records from May and early June (as

Banded Hairstreak *(Satyrium calanus)* *(continued)*

late as June 11). Patterson reported by email a May 2016 sighting in the NAT Kisatchie unit (not reflected in the distribution map). In the Acadiana region, three were seen at Acadiana Nature Park (early May). Brou (2009a) caught over one hundred at light traps in the FL Parishes during late Apr and May, and Trahan (2009b) also reported one, photographed at a moth light in north LA. Other southeast LA records include one Pearl River count (June) and at Asphodel Plantation (May and June).

Host Plants: This hairstreak has been reported to use boxelder, black maple, chestnut, crabapple, hickory, oak, black walnut, willow, and ash (Carolina, green, white) in LA.

King's Hairstreak (*Satyrium kingi*)

ventral

Description and Behavior: King's Hairstreaks resemble several other *Satyrium*, including Banded, Striped, Hickory, and Edwards hairstreaks. The latter two have not yet been found in LA. King's are larger than Banded and Striped hairstreaks. All three can be found in the same immediate habitats. In fact, on three separate occasions I have seen King's and Striped on the same day, within sight of each other. In LA, Banded Hairstreaks are typically a darker shade of gray than King's. The primary diagnostic distinction between Banded and King's hairstreaks is the red cap over the blue spot on the King's ventral hindwing. Also, the female Banded has a red or orange spot over the tail dorsally. While the Striped hairstreak is the same shade of gray as the King's and has the red cap over the eyespot, the multiple ventral stripes make it easy to identify a Striped.

I have found the King's flight to be less "swirling" and slower than both Banded and Striped hairstreaks. They like to dash about (as only a hairstreak can "dash") in the shady understory of deciduous woods. They land on the tips

King's Hairstreak (*Satyrium kingi*) (*continued*)

ventral

of outreaching branches of their food plant, primarily in spots of sunlight, but are wary and difficult to approach. I have never seen this hairstreak taking nectar, possibly because I have never seen it in an area where nectar sources are present.

My Records: I have found this hairstreak flying in suitable habitat at the Blue Hole Rec Area, Longleaf Vista Rec Area, and Kisatchie Hills Wilderness Area. I've also seen multiple specimens in the Alexander SF, next to Indian Creek Rec Area, flying from mid-May to mid-June. Most recently, I found a colony at Lake Ramsey Savannah WMA in late May 2015. All of these locations were deep within hardwood bottoms near water, and all had extensive stands of sweetleaf.

Distribution and Abundance: In May 2016, Patton posted a photograph on the LA listserv of one at Chicot SP, and in late May 2017, I found a ragged female maybe a mile away at the LA Arboretum. Brou (2008b) reported taking eighty-one high-quality specimens over twenty-five years at light traps near Abita Springs, along with visual sightings of others (tattered specimens were not recorded), primarily found near sweetleaf present at that study site. They were

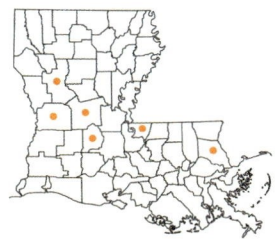

found from early May until mid-June, most common during the first week of June. This hairstreak was also reported by Israel near Rogillioville in mid-June. It has been recorded in only six parishes, and was listed in 2015 as a Tier III, SU-ranked SGCN in LA.

Host Plant: This hairstreak's distribution closely mirrors that of its larval host plant (probably its only host plant), sweetleaf. Sweetleaf is a shrub or small tree with a short trunk, up to twenty feet tall. It ranges along the Gulf Coast from central FL to east TX. The leaves have a sweet taste to livestock; thus the other common name, "horse sugar." Its habitat is indicated as moist valley soils in the understory of hardwood forests.

GOSSAMER-WINGED BUTTERFLIES (Lycaenidae)

Striped Hairstreak (*Satyrium liparops*)

ventral

ventral

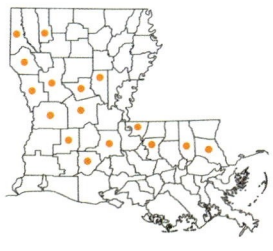

Description and Behavior: I have never seen this hairstreak in numbers, but always as singles. All of my sightings in LA were in the same immediate area as colonies of King's Hairstreaks and flying at same time. Several were seen along slow-moving (if moving at all) creeks, actually flying along with the King's. Like King's, as it flies in the understory of the forest, it seems to like to perch in the sun. Also like the King's, it appears to be single-brooded, flying from late Apr into the third week of June. Regarding the subspecies in LA, none of my specimens from this state have the orange patches of the FL subspecies (*S. l. floridensis*). Rather, they appear to either be the nominate *S. l. liparops* or *S. l. strigosum*. Brou (2007b) noted that only two of his seventy-six specimens possessed the dorsal mid-forewing patch.

My Records: This hairstreak's preferred habitat is typically moist, deciduous thickets and swamps located within central LA's piney woods. I have found the Striped Hairstreak in the VER, RAP, and NAT units of Kisatchie NF. I've yet to find it, for example, in the Acadiana or southwest LA regions, where pine forests are not the dominant habitat.

Distribution and Abundance: Records for this hairstreak are from across the state but sporadic. In north and central LA, those records include SAB (mid-May), LAS (June), DES (late May), and GRA (late Apr), all found in hardwood forests. Records from the FL Parishes also fit this mold and include Tunica Hills WMA and Greenwood Park (both early May). There are three records from the Cajun Prairie region in May and June. It is mostly rare across its LA range but, near Abita Springs, Brou (2007b) has reported finding seventy-six adults in ultraviolet light traps, with others observed over a twenty-five-year period.

Host Plants: This hairstreak uses many plants, including ash (Carolina, green, white), river birch, highbush blueberry, black cherry, chestnut, chokeberry (red), crabapple, hawthorn, hickory, American hornbeam, oak, pear, pecan, plum, wild azaleas, huckleberry, and serviceberry. In AL, sparkleberry is listed as both a larval and an adult food plant.

Red-banded Hairstreak (*Calycopis cecrops*)

ventral

ventral

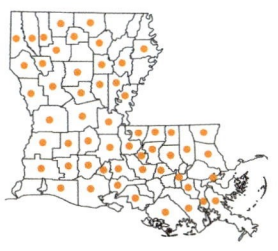

Description and Behavior: This hairstreak likes wooded regions, and I've found it deep in the woods, including pine forests. It flies low, regularly landing among the dead leaves on the forest floor. When disturbed, it may fly up and land on a high branch or stay low, disappearing in the debris. It will venture out into open fields and disturbed areas to take nectar. It is easily watched at flowers and seems particularly fond of white flowers such as boneset and sumac. While it does "perch" like other hairstreaks, I don't find that it will defend perches as other hairstreaks will do. I don't recall ever seeing it engaged in what I have described as aerial dogfights, but it will chase other Red-banded Hairstreaks for a short distance. Mostly, I see it when it suddenly flies up as I walk along, only to move a few feet and then land just as suddenly. Although this butterfly is one of our smaller hairstreaks, the blue on its upper dorsal wings is visible while it is in flight.

My Records: I consider this to be LA's most common hairstreak, and I have found it in every region of the state. My records suggest it flies from late Feb (during mild winters) until Nov. Unlike the majority of our hairstreaks, which have only one brood, it is multi-brooded, and can be seen flying in numbers.

Distribution and Abundance: Composite records show this hairstreak is common around the state, recorded during every month of the year. In north LA, it has been recorded in CAD from late Feb until late Nov, most common in Mar and Apr. It was the most common hairstreak in the Cajun Prairie during the survey of that region, flying from Mar until Oct. Per Israel (1981), it is locally common from late Jan or early Feb into early Dec in the Felicianas, "best caught at flowers where it is constantly taking off and then landing again head down often on the underside of leaves of shrubs." Brou (1974) reported taking about seven hundred specimens at light traps over the previous twenty-two years from SJB and STA. His records reflect a flight season from late Feb (with a record in mid-Jan) into late Nov over four broods (late Feb/Mar, early June, late July/early Aug, and Oct). Records from southernmost LA also extend from Feb to Nov.

Host Plants: In LA, it is reported to use shining sumac, wax myrtle, and oak trees. It was found once on goatweed in WFE. In southeast TX, it reportedly also uses crotons and oaks. In south FL, it uses Brazilian pepper.

GOSSAMER-WINGED BUTTERFLIES (Lycaenidae)

Dusky-blue Groundstreak (*Calycopis isobeon*)

ventral

Description and Behavior: NABA did not include the Dusky-blue Groundstreak as part of LA's fauna; however, BAMONA did. I found only one other source (Scott 1986) suggesting that *isobeon* might occasionally migrate into LA and MS. Traditionally, this species is a denizen of the woods and shade, primarily found inside the woods, along tree lines, or in open areas only a short distance from the trees. Like its close cousin, *C. cecrops*, this hairstreak flies low, typically below knee level.

The distinguishing features between *isobeon* and *cecrops* are geography and physical description. In terms of geography, *isobeon* has been historically reported in TX and then south into Mexico and beyond, while *cecrops* was found throughout the southeast US extending to east TX. There are two primary distinguishing physical features between these two hairstreaks. First, *isobeon* typically has a very thin red-orange line on the ventral forewing, while in *cecrops* that line is wide. Second, on *isobeon*, the larger eyespot between the two tails on the ventral hindwing is more red-orange than black or at least equal parts of each, while in *cecrops* there is basically no red-orange. Further, some sources

ventral

suggest that in *isobeon* the red bleeds past black and white lines into the tail area on the hindwing.

There is not clear agreement that *C. isobeon* and *C. cecrops* are really separate. Both occur in the Houston area, where some specimens appear to be intergrades showing the narrow red post-median line on the ventral forewing, but only a small red cap on the large black spot. The Tvetens (1996) have suggested, "Certainly there seems to be a gradual transition between typical forms through the Houston area, and clear identification awaits more detailed taxonomic work." Several sources suggest the two species interbreed in those areas where their ranges overlap. Others wonder whether the populations in this region of overlap are evidence of clinal intermediates or hybridization between the two species or if they constitute intergrades between the two, yielding proof of only one species. Further investigation is needed.

GOSSAMER-WINGED BUTTERFLIES (Lycaenidae)

Dusky-blue Groundstreak (*Calycopis isobeon*) (*continued*)

My Records: Before 2012, I had seen this hairstreak only in TX with the closest location to LA being Brazos Bend SP, south of Houston. In 2012, I was in CAM in mid-Oct and saw, briefly, what I identified as a Red-banded Hairstreak. Roever verbally reported to me that he was in the same area the next weekend and believed what he saw were not Red-banded Hairstreaks but Dusky-blue Groundstreaks. So I returned to the referenced area of west CAM in early Nov 2012 and found at least one specimen that strongly matched the distinctive markings identified for this hairstreak. The other two I was able to inspect closely were so badly faded and/or tattered that I could not make a determination.

In July of 2013, I saw several specimens in LAFA that much more closely resembled *isobeon* than *cecrops*. Coincidentally, there were several plants in the immediate area that were transplanted, soil and all, from CAD, where Dusky-blue Groundstreaks had already been documented by BAMONA. I sent pictures to several people and received divided opinions with most agreeing that at least one, probably two, were Dusky-blue Groundstreaks. In late Oct 2014, while at Freshwater City Lock Rec Area, in habitat identical to that found in lower CAM, I found another of what I considered to be *isobeon*.

I believe that, while not perfect matches to those typically seen in the Rio Grande Valley, what I have found in CAM and VRM are Dusky-blue Groundstreaks. I found more specimens of what I believe to be this species in both western CAM and lower VRM in Oct 2016. This hairstreak had been previously identified in CAM (as well as immediately across the state line in Jefferson County, TX), so it would not be unrealistic to find it there again. If my diagnosis is correct, *isobeon*'s range in LA should be expanded to five parishes, as far north as CAD, and as far east as LAFA.

Distribution and Abundance: The first report (BAMONA) was of a sighting in CAM in mid-Dec 2007. Within a year, a second sighting was reported in CAD (Shreveport) in late Sept 2008. A third showed up in that parish in 2011.

Host Plants: The caterpillars of *C. isobeon* and *C. cecrops* look identical. Several sources indicate the caterpillars of both species eat rotting and decaying leaves, fruits, and other detritus on the ground under trees.

Gray Hairstreak (*Strymon melinus*)

ventral

dorsal

GOSSAMER-WINGED BUTTERFLIES (Lycaenidae)

Gray Hairstreak (*Strymon melinus*) (*continued*)

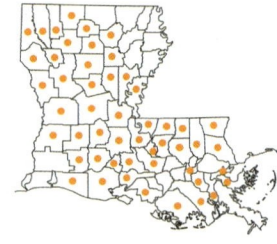

Description and Behavior: This hairstreak does not appear to be habitat restricted. It typically prefers more open, disturbed areas and is readily attracted to multiple species of flowers. It loves lantana and will visit gardens where that plant is present. Its flight is fast and erratic; once disturbed, it can be so fast it is hard to follow, only to finally return whence it came, or nearby. When at flowers, it can be approached slowly and netted or photographed.

In late summer and into the fall, the females can be quite large and might be confused with the White M Hairstreak; however, this hairstreak is ventrally a brighter shade of gray than the White M and has two red anal spots as opposed to one large one. Males of this species have not only the red hindwing spots, but also red abdomens. Finally, White M hairstreaks are blue dorsally while this hairstreak is dark gray with a red or orange anal spot above the tail.

My Records: This is probably the second most common hairstreak in LA. It is multiple-brooded from late Feb into early Nov (early Dec once). At times several can be seen in one day, particularly in CAM. For example, I saw sixteen in the western portion of that parish in July 2013. Elsewhere, in July 2013, I saw forty-five in the open glades at Copenhagen Hills.

Distribution and Abundance: Also known as the "Common Hairstreak," it has been recorded throughout the state. In CAD, it will fly from Feb through mid-Nov. In the Cajun Prairie region, records were from Apr through Oct. It is locally common in both of the Felicianas with records from late Mar into early Nov, best found at flowers. Eight turned up in Brou's (1974) ultraviolet light traps in SJB. In southernmost LA, it has been recorded from Mar through Dec (December 12 at Grand Isle).

Host Plants: In LA, it uses croton, red clover, members of the legume and mallow families, hibiscus, and oak. On one occasion, this hairstreak's caterpillars were found on corn in BOS.

Mallow Scrub-Hairstreak *(Strymon istapa)*

ventral

ventral

GOSSAMER-WINGED BUTTERFLIES (Lycaenidae)

Mallow Scrub-Hairstreak (*Strymon istapa*)   (*continued*)

Description and Behavior: The range of this south TX hairstreak extends into the Corpus Christi and Lavaca Bay areas. BAMONA has listed it as present in the Houston and Baytown areas. It is a smallish, gray-brown hairstreak with a single, short tail similar to that on an Eastern Tailed-Blue. It has a small anal eyespot just above the tail, capped slightly in red or orange. There are two or three small black spots along the leading edge of the ventral hindwing.

My Records: Until Oct 2012, my experience with it has been in south TX; however, in mid-Oct 2012, I found a fresh male in west CAM, north of Holly Beach. As its name would suggest, the hairstreak was found in the immediate vicinity of some mallow plants, fluttering low to the ground.

Distribution and Abundance: There have been two other sightings in LA, in the extreme northwest corner of the state near the TX border. The first was in Shreveport, and the second was at Corney Lake Unit. Both were in Nov 2007.

Host Plants: Plants in the mallow family, including dollarweed in central TX.

White M Hairstreak *(Parrhasius m-album)*

ventral

Description and Behavior: This is a large hairstreak. While smaller than the Great Purple, it is larger than all others except possibly female King's Hairstreaks. The females are slightly larger than the males. Ventrally, the two sexes look the same. Dorsally, they present a spectacular shade of bright, metallic blue. The females are a darker blue and have a broader black border that can cover more than half of the dorsal forewing. Most of the specimens I have seen in LAFA were males, perched in the afternoon on the tips of several kinds of bushes and weeds at about waist height, both in the sun and shade. They will chase each other and other butterflies, dashing off with a sudden flash of bright blue, only to return a few seconds later.

My Records: This hairstreak is typically rare, but I have consistently seen specimens each summer since 2003 at one location in LAFA. Records for this site

White M Hairstreak (*Parrhasius m-album*) (continued)

ventral

extend from late May into early Aug, with multiple butterflies seen over that period. Other than the one spot in LAFA, I never know when it will be seen. It appears somewhat regularly at Thistlethwaite WMA. Once, in May, I saw eleven flying along a deer hunter's trail, perching in the sunlight on high limb tips. I've seen it a few times on the forest floor in deep woods (LAFA, SMN, and DES). It is attracted to flowers and, in the fall, seems to prefer white flowers such as boneset. I have also seen it at the Catahoula Butterfly Garden in late Mar, where there are multiple blooming trees and bushes with bunches of small white flowers attractive to several kinds of hairstreaks, including this one.

Distribution and Abundance: There are records from CAD for early Mar into Apr, then sporadic sightings from June to Nov. Only one was seen in the Cajun Prairie region (LAFA, June). Patterson reported by email a May 2016 sighting in the NAT Kisatchie unit (not reflected in the distribution map). A male was seen and photographed during the 2017 NABA count at Allen Acres in VER Parish in late July (also not reflected in the distribution map). In mid-Feb 2015,

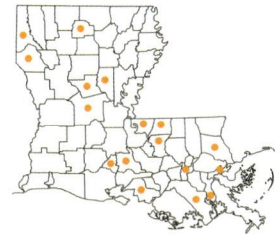

another was seen in LAFA during an extended warm stretch. Two sightings have been recorded at Peveto Woods in mid-Oct. In southeast LA, Brou (2000) suggested it has up to six broods annually in the FL Parishes, flying from as early as late Feb/early Mar through early Nov. Verbally, Brou indicated he also had records in Jan and Dec, with over one hundred collected using ultraviolet light traps. It is rare in EFE in Mar to Apr and then June to Nov. It is also rare in WFE, with records of three broods extending over periods from late Mar into early Apr and then late June to Nov (Israel 1981). It was recorded twice, both as singles, on the Pearl River count. It has also shown up occasionally on the New Orleans count (June). In southernmost LA, it has been recorded at Galliano and Golden Meadow (Apr, June, and Sept) and Grand Isle (Aug). The Golden Meadow sighting was at an ultraviolet light trap.

Host Plants: LA sources have basswood and oak as host plants. In the South in general, hosts are reported to be just about any species of the oak family, but especially trees with narrow or very lobed leaves.

GOSSAMER-WINGED BUTTERFLIES (Lycaenidae)

Cassius Blue (*Leptotes cassius*)

ventral

dorsal

Description and Behavior: The Cassius Blue is present throughout the year in south FL and south TX. It has been recorded on a few occasions in MS and AL. This blue and the Marine Blue have been characterized by several authors as being "zebra striped," ventrally. I have always thought the striping was darker on this blue. Dorsally, it shows a pale blue color. This blue is typically smaller than the Marine Blue and ventrally shows more white in the submarginal region.

My Records: I have seen it only in FL and TX.

Distribution and Abundance: The first LA record was in the NO area, where Brou (2013b) reported taking it during the summer months of 1963 and 1964 (a new state record at that time). Ironically, he advised via e-mail that "it was the only blue in New Orleans back then and it was common on what is now a high school campus and on the campus of the University of New Orleans." It was next reported in late June 1968, in SHE. It showed up again in the NO region, in Oct of 2000, at which time several were found on plumbago. Ross and Weldon (2003) described this blue as an "occasional colonist in New Orleans," with a flight period from July to Aug. They felt it was most likely transported here from FL on blue-flowering plumbago nursery stock, its caterpillar food plant. It is also occasionally present in the Houston area in late summer and fall, so it might show up in the southwest corner of LA.

Host Plants: Its primary food plant is plumbago. In south FL, it uses herbs and vines of the bean and leadwort families, such as milk peas. In southeast TX, string and lima beans are also used as well as rattlebox.

GOSSAMER-WINGED BUTTERFLIES (Lycaenidae)

Marine Blue (*Leptotes marina*)

ventral

ventral

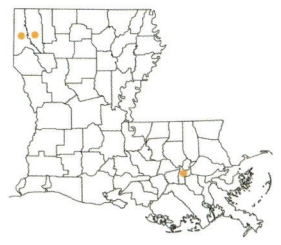

Description and Behavior: This blue is ventrally somewhat similar to the Cassius Blue. Both of these blues have two very noticeable submarginal black spots on the hindwing; however, on the Marine Blue those spots are rimmed with metallic, silvery blue scaling. On the Cassius, those spots are typically circled by an orange-reddish ring. Subject to northward migration out of south TX, it has moved into the Houston region as early as Apr, remaining through the summer into Oct or Nov. It has also been reported in Galveston County. To the northwest, it has been recorded at Red Slough WMA in southeast OK. To the east, I found three old records in MS, including one from the extreme southwest section of that state.

My Records: I have found this blue primarily in south TX. I found one in Shelby County, TN, in June, which seemed to be an odd time of the year for one to have found its way that far north.

Distribution and Abundance: I found five records in LA. The earliest was in late Sept 2006, at Shreveport. The second was also in Shreveport in early Oct of that same year. In mid-Nov the following year, a third was reported at Cane's Landing. The fourth was reported by Ronnie Gaubert on his website, "Butterflies of Louisiana Photo Gallery," on which he had posted a picture of the blue he found near Ruddock (in SJB) without date. And Patterson reported by email a June 2004 sighting in the NAT Kisatchie unit (not reflected in the distribution map).

Host Plants: In southeast TX, it reportedly uses numerous legumes and leadwort.

GOSSAMER-WINGED BUTTERFLIES (Lycaenidae)

Western Pygmy Blue (*Brephidium exilis*)

ventral

dorsal

Description and Behavior: Some sources consider the Western Pygmy Blue and the Eastern Pygmy Blue to be one species with *pseudofea* a subspecies of *exilis*; however, the prevailing view is that the two are separate species based on genitalic differences. Further, this blue uses a wider variety of larval food plants and can survive in more varied habitats than *pseudofea*. Its range extends from west LA along the Gulf Coast to south TX, and then westward. Found in coastal marshes, dunes, along roadsides, and in waste places (all present in CAM), this blue is extremely small and can easily go unnoticed. It flies weakly, low to the ground, never straying very far from its food plant or from the coastal marshes and/or dunes where the food plant grows.

Neither this "blue," nor its cousin, are particularly blue; rather, they are almost all brown. Both sexes are dorsally brown with some blue showing at the wing base. Ventrally, they are whitish gray at the base of both wings and white on the fringes. In contrast, the Eastern Pygmy Blue is completely brown ventrally, including the fringes.

My Records: Before 2011, I had never seen this butterfly within the state. In Aug of that year, I saw it flying near Rockefeller NWR in CAM. Later that year, in Oct, I saw several in the western portion of the parish, both at Peveto Woods and Sabine NWR. In 2012, I found it at Holly Beach.

Distribution and Abundance: This tiny butterfly was first reported as common in Sept 1970, found near the East Jetty area of CAM. Other records include Broussard Beach (Aug and Oct), Rutherford Beach, Mae Beach, and Holly Beach (all in Oct). In the latter three locations, it was present in good numbers in the grass along the beach. The latest recorded date was Nov 2, 2013, in that same area.

In the early twenty-first century, two major hurricanes have hit and significantly affected CAM. The pygmy blues I saw there in 2011 and 2012 were all unmistakably *exilis*. I found the Western species to also be very common jsut across the TX state line at, for example, Baytown, and I believe *exilis* is repopulating CAM from the west. It was listed in 2015 as a Tier II, S1/S2-ranked SGCN in LA.

Host Plants: In southeast TX, it is reported to use "saltbush, pigweed, glasswort and other salt- and alkali-tolerant plants," including Virginia glasswort, horse purslane, and seepweed. In west CAM, it is believed to use American glasswort and southern sea-blite. These plants grow in salty conditions along ocean beaches.

GOSSAMER-WINGED BUTTERFLIES (Lycaenidae)

Eastern Pygmy Blue *(Brephidium pseudofea)*

ventral

dorsal

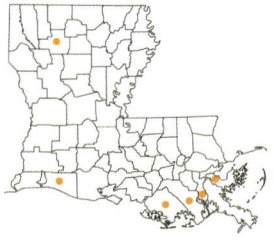

Description and Behavior: Like its Western cousin, this is a tiny butterfly, approximately one-half inch from wingtip to wingtip. It flies inches above its food plant. This little blue is habitat restricted, reported to prefer hypersaline salt flats where succulent plants like perennial glasswort grow. In some areas of its range, the caterpillars are tended by ants, and the existence of this symbiotic relationship might limit its distribution. In contrast to the Western Pygmy Blue, which has been found only in extreme southwest LA, this butterfly is more common along the Gulf Coast in southeast LA, possibly as a result of its migration up from FL along the Gulf Coast.

My Records: I have seen this blue only in south FL.

Distribution and Abundance: The first report of this blue in LA was on Isle au Pitre, off the southeast corner of the state in June 1969. It was found in abundance on the Chandeleur Islands in July and Aug 1969. Strickland and Strickland (2005) generated a website that included three pages on the "Butterflies of the Genus *Brephidium* (Pygmy-blues)," which addressed the overlapping and conflicting ranges of this species and the Western Pygmy Blue within the state. As part of that website, the Stricklands reported that in 1980 they found this species on Grand Isle, about eighty miles east of the Chandeleur Islands. In mid-July 1992, they returned to the East Jetty location in CAM where they had previously caught *exilis* only to find what they believed were *pseudofea*. The Stricklands reported that the 1992 specimens suggested the potential of some hybridization between the two in this area. They felt that, between their first and last visits to CAM, there has been an expansion westward of *psuedofea*. In late Oct 2016, Patton reported finding this species on the LA listserv (with pictures) at Broussard Beach in CAM.

This species was also reported on the 2003 and 2004 Grand Isle counts (forty recorded in 2003, found around its food plant on the inland side of the island). Other locations include Ile D'Canoe and Port Fourchon (Mar to June, Aug to Sept, and Nov), Cheniere Caminada, Grand Isle (Apr to Oct and Dec), and Dulac. It was listed in 2015 as a Tier II, S1/S2-ranked SGCN in LA.

Host Plant: It was found in association with glasswort during the 2003 Grand Isle count. Perennial glasswort has been identified as the host plant used in the northeast section of FL.

Eastern Tailed-Blue (*Cupido comyntas*)

ventral

dorsal female

dorsal male

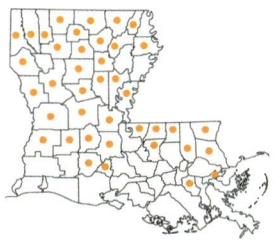

Description and Behavior: This is LA's only tailed blue. This butterfly is dimorphic. The males are blue dorsally (but not as bright blue as the azures) while the females are slate gray. They look the same ventrally, but the females are typically larger. It is multi-brooded. My earliest record is Mar 3 and latest record is Dec 6 (in EBR). Both sexes visit low-growing, small flowers readily, preferring white and blue. It does not seem to fly in areas where the grass is tall. It has also been seen at mud.

My Records: I have seen this blue occasionally in Acadiana (for example, only four times in LAFA over the years) and much more frequently in central LA. It is occasionally present at Thistlethwaite WMA, primarily in the main pipeline cut that runs through the middle of that unit. I have found it along the levee road at Indian Bayou WMA. It is more common at the GRA unit of Kisatchie NF, flying in the grass trails that cross Stuart Lake Road. It is also present in the low grass along the roads of the NAT unit and in the power-line cut at the RAP unit. I once found it to be abundant at Cooter's Bog (July). It flies in the open glades at Copenhagen Hills (June and July, with twenty-six seen on the latter visit). In southeast LA, I found it at Sandy Hollow WMA (Sept).

Distribution and Abundance: This blue has been found throughout the state, has been recorded in every month of the year, and has been reported as abundant during the spring and summer. There are records from CAD in early Mar until late Oct, with it being common to abundant during Apr and May. It was seen in small numbers during the Cajun Prairie survey during Mar, May, June, Sept, and Oct. During the 2002 Cameron count (Aug), twenty-two were reported. Records suggest three broods in the Felicianas, where it is common, flying from Mar into Dec in open, cleared areas. It was reported on one New Orleans count (1999) and one Pearl River count (2007).

Host Plants: It is reported to use vetch here in LA. In southeast TX, it reportedly uses numerous herbaceous plants in the Fabaceae family.

Spring Azure *(Celastrina ladon)*

ventral

dorsal male

ventral

Description and Behavior: In 1884, William Henry Edwards described all of the Spring Azure's forms as a single, polymorphic species. Even after three newly recognized species were segregated out of the *ladon* complex, there were still as many as five "sibling species" within that complex, including the Spring and Summer Azures. Older references described it as *Lycaenopsis argiolus*. Pelham (2008) used *C. ladon,* and I have followed his lead.

The majority of the state reports for this species were recorded before the Spring and Summer azures were separated. As a result, other than by date, it is difficult to identify which sightings were for which species. Early spring dates would suggest this species but, by mid-May, I look more to the ventral appearance as a guide. Specifically, the Spring Azure appears more prominently marked around the wing margins and has more and darker spots and dashes inside those margins. Several sources suggest this species and the Summer Azure can be differentiated by examining the upper forewing scales of male specimens through the use of a hand lens. Specifically, they suggest the presence of long, "translucent" scales on the forewing of males of this species that overlap the blue scales, causing a blurred or fuzzy appearance. Those long scales are not present on males of the Summer Azure, giving the blue scales a more focused appearance.

GOSSAMER-WINGED BUTTERFLIES (Lycaenidae)

Spring Azure (*Celastrina ladon*) (*continued*)

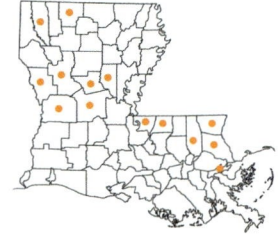

My Records: One of the more likely spots to see the Spring Azure is along the main road through the Weyanoke area in late Mar. I saw four one day in mid-Mar, on Castor Plunge Road in the RAP unit of Kisatchie.

Distribution and Abundance: This blue was initially identified in LA as *C. argiolus pseudargiolus.* It was considered to be rare, with a few specimens taken in the pine flats of the FL Parishes during May. Since that report in 1950, additional sightings, using different names, have been reported from around the state. In north LA, Kendrick-Lacy advised via e-mail that she saw azures in CLA, extending from spring throughout the summer. Trahan's records (2009a) reflect *C. ladon/neglecta* as occasional in CAD from late Feb into mid-Mar, and then again from mid-Apr to mid-Oct. Using that same generic name, two were reported as seen during the study of the Cajun Prairie region, one in Apr and one in June. All of these records appear to reflect both species, not just the Spring Azure. Jonathan Clark reported, in a post to the LA listserv (with a photograph), a Spring Azure in LAS in Feb.

Composite records indicate a more common presence along and then east of the Mississippi River. These records include the Baton Rouge and NO areas along the Mississippi River with dates in Mar, Apr, and May, near forest cover. As part of his Tunica Hills survey, Israel (1981) reported four broods from late Feb into mid-Dec. He noted that the Spring Azure was locally common, flying "high and fast along forest edges," and that it was seen at flowers, mud, and dung. At his home in EFE, he recorded it as uncommon from Jan to Dec and most common in May and June—which, by date, would actually be the Summer Azure. Other records in the FL Parishes include near Abita Springs (Feb and Mar), Bogue Chitto SP (Apr), and in TAN (Apr).

Host Plants: Several larval food plants have been identified for the Spring and Summer Azures, including highbush blueberry, red buckeye, coral bean, black cherry, crabapple, devil's walkingstick, dogwood, oak, plum (American and Mexican), privet, viburnum, and witch hazel.

Summer Azure *(Celastrina neglecta)*

ventral

dorsal male

GOSSAMER-WINGED BUTTERFLIES (Lycaenidae)

Summer Azure (*Celastrina neglecta*) (*continued*)

dorsal female

Description and Behavior: In its 2001 checklist, NABA noted the existence of studies suggesting that the summer populations of the Spring Azure were a separate species. After referencing additional studies, NABA concluded in its 2008 *4th of July Butterfly Count Reports* that the data "did not conclusively establish that *C. ladon neglecta* is genetically isolated from spring-flying populations." In contrast, Pelham (2008) gave it full species status, and Glassberg (2012) reported that unpublished DNA data grouped "all spring populations as one species and summer and fall populations as another." I have followed Pelham and Glassberg.

This blue is typically seen as singles, flying high along a tree line or forest's edge. Azures are easy to differentiate from the other blues seen in LA by the height of their flight. The state's other blues are denizens of open, disturbed areas and fly low along the grass tops. In contrast, azures inhabit more wooded areas. Also, while the flight of the state's other blues is slow and fluttery, weav-

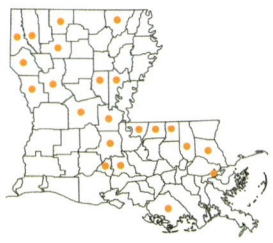

ing around and through the grass, both of the azures have a much faster and more direct flight. Finally, the other blues will regularly perch on grass tips as well as at small, low-growing flowers, but the azures perch several feet off the ground on extended leaf tips or at flowering trees like redbuds.

My Records: My records suggest the Summer Azure is seen more often than its spring cousin in the Acadiana region. These sightings include Indian Bayou WMA, Thistlethwaite WMA, the St. Landry area in AVO, and Indian Creek Rec Area. A good place to closely observe Summer Azures is Clark Creek Nature Area in Pond, MS, just one mile above the LA state line in WFE. The azures come down to the wet sand in the creek bed and are easily viewed. In early June 2013, I saw two at Sicily Island Hills WMA, which has terrain and habitat very similar to Clark Creek Nature Area. The year 2017 was a boom year for this species. A male was documented during the 2017 NABA count at Allen Acres in VER in late July, and I saw another at Valentine Lake in RAP in mid-Aug. Also, Patton posted pictures of several on the LA listserv from the Pine/WAS area in early Aug (the late VER and WAS records are not reflected in the distribution map).

Distribution and Abundance: Given the extended period over which this azure flies, I believe there is more than one brood. Israel (1981) reported four broods for *C. a. pseudargiolus* in WFE, of which the last two should be this species. Sightings reported during June in WFE and SHE (near Greensburg) would, based on the dates, probably be this species. Several specimens in the LSU State Collection taken in Aug in WEB, EFE, and WAS should also be this species.

Based on the dates, I would propose that those recorded during the 1997 Barataria count (June), the 2000 Alexandria count (Aug), the 2004 Global Wildlife count (Aug), and the 2002 New Orleans count (June) be considered as this species. Other locations include Bodcau WMA and Delaney Hills (late May through Aug) and Roy Quad (June). It has also been reported at Eddie Jones Park (late Oct 2012), and from southernmost LA in TER (July 2007).

Host Plants: See the generic list provided in the Spring Azure account. In AL, swamp dogwood and sourwood are reported as used early in the flight season, with wingstems and tick trefoils used later in the season.

GOSSAMER-WINGED BUTTERFLIES (Lycaenidae)

Ceraunus Blue (*Hemiargus ceraunus*)

ventral

dorsal female

dorsal male

Description and Behavior: This is a common blue in
south TX and south FL. It has occurred irregularly
in the Houston area, although not as commonly as
Reakirt's Blue. It has also been reported in Galveston
County. It is about the same size as the male Eastern
Tailed Blue and flies in a similar fashion, low to the
ground and just above the grass. I once found it fly-

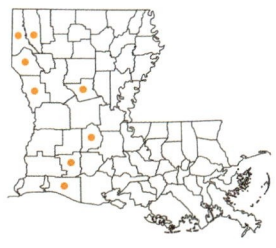

ing in east TX in several ditches filled with frog fruit. It has shown up in LA several times, primarily in west LA, suggesting they migrate occasionally from TX.

My Records: During the 2012 Duralde count (Sept), I found a lone male. Also,
during Oct 2012, on consecutive weekends, I found two fresh males near Grand
Chenier, and then three males in west CAM. This blue was still flying at Holly
Beach in early Nov 2012. I did not see it again until Sept 2013, at Holly Beach.

Distribution and Abundance: Prior to 2012, there were four records, all from
CAD and all in Aug and Sept. In 2012 it was first reported in Sept and Oct, including multiple sightings in CAD (Eddie Jones Park). It was also seen in that
region at Bodcau WMA (twenty-two seen in Oct) and in DES (Oct), making
2012 a banner year for this blue. It showed up again in good numbers at Bodcau
in Aug 2016. In the central LA region, one was recorded at the Catahoula count
in Oct. Despite a hard winter with at least two freezes, this blue was seen again
in Oct and Nov 2013. Specifically, in mid-Oct, more than twenty were seen at
Holly Beach and Constance Beach. A few days later, fifty-two were recorded
at Mae Beach.

Host Plants: In southeast TX, it reportedly uses numerous herbaceous and
woody legumes. In south FL, it uses herbs and vines of the bean family like
Alicia, milk pea, partridge pea, and indigos.

GOSSAMER-WINGED BUTTERFLIES (Lycaenidae)

Reakirt's Blue (*Echinargus isola*)

ventral

dorsal female

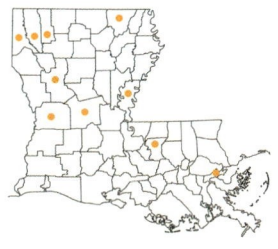

Description and Behavior: Typically found in TX and west thereof, it has appeared in the Houston area as early as Mar or Apr, flying until Nov, most abundant in June. It has also been reported in Galveston County. To the north, it is listed as a member of the butterfly fauna at Red Slough WMA in southeast OK. Its preferred habitat is open, weedy disturbed areas and roadsides. This blue and the Ceraunus Blue are superficially similar, but a row of white-rimmed black spots on the ventral forewing of this blue is the primary distinguishing marking between the two. None of the other blues that might be encountered in LA share this distinctive marking.

My Records: My only sighting within the state was at the Claiborne Multi-Use Area in RAP in June 2014.

Distribution and Abundance: This species was first reported in LA in 1954 in CON and MOR, at that time thought to be the eastern edge of this species's range. Most subsequent sightings have been from the western side of the state. In north LA, it has shown up three times in CAD (May and twice in Nov). Other sightings from that portion of the state include Corney Lake Unit (late May), Bodcau WMA (early June, three seen), and in WEB (mid-June). In the central LA region, it has been recorded several times in Kisatchie NF in NAT (Apr, May, and June, including eight on one occasion). In June 2014, one was seen at Cooter's Bog. Not all records are along LA's border with TX. For example, one blue was reported (with a picture) during the 2001 New Orleans count. It was also found at Greenwood Park in EBR in mid-Nov. These sighting would seem to expand the eastern limit of this species's range beyond the TX border.

Host Plants: Listed as using mimosa and honey mesquite in LA, it has also been found on goat's rue, a legume, in VER. In southeast TX, it reportedly uses numerous legumes in the Fabaceae family.

METALMARKS (Riodinidae)

Little Metalmark *(Calephelis virginiensis)*

ventral

dorsal

Description and Behavior: This small butterfly (LA's only metalmark) can easily be missed. Its flight is fluttery and low. It doesn't move around much unless disturbed, and will regularly perch on the underside of grass or leaves close to the ground. The best way to locate Little Metalmarks is to find areas of suitable habitat (open, pine flats) with large stands of its larval host plant. While never far from its food plant, the Little Metalmark does not typically perch on the thistle; rather, it prefers to perch on plants about six to twelve inches in height in the immediate area around thistle. At Cooter's Bog, Little Metalmarks are present upland from the bog, in an open, grassy area inside the pine forest. In the morning, the males perch on tall grass and don't move much unless disturbed. As the day warms, both sexes will seek nectar, preferably at yellow flowers, including members of the black-eyed Susan family. I twice saw males chasing each other.

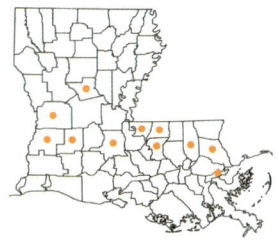

My Records: My initial experience with this little jewel was in the two Felicianas in late Oct and early Nov. In mid-Oct, I saw over ten in less than thirty minutes walking a power-line cut across the highway from Asphodel Plantation. I have seen it at Mary Ann Brown Preserve in the last weekend of Oct and the first weekend of Nov. In mid-May 2013, I found one at Cooter's Bog. In early July of the same year, I found ten there and then three in late Aug. The majority of the Cooter's Bog area was burned over the 2013–14 winter, but by late Aug 2014, the metalmarks had moved back into the area. In 2017, I found one at CC Road Savanna Preserve in ALL (mid-Mar) and another at Drake's Creek Natural Area in VER (also mid-Mar).

Distribution and Abundance: This metalmark has never been widely reported. Most records are from the FL Parishes, including the Greensburg area (Apr and June). The LSU State Collection contains specimens from EBR, taken in early Sept and Oct along the River Road. This butterfly can be common around Asphodel Plantation. Israel verbally reported that he had seen this butterfly as early as the first week of Mar and as late as the first week of Nov. He observed that the two main broods fly during the second half of July and the second half of Oct. In southeastern LA, three turned up during the 1996 New Orleans count (June), but not since. It has also been reported in TAN (Mar).

West of the Mississippi River, two specimens were reported near Pollock in Apr 1964. One was recorded in Nov as part of the study of the Cajun Prairie region, sighted at the ALL-BEAU line along Hwy 190. In VER, it was reported on two Fort Polk counts (Apr and Sept) and nearby at Dove Field Rec Area (June). Elsewhere, there is an old record (1931) from Opelousas, although the habitat in that parish does not match the typical habitat for this butterfly. It was listed in 2015 as a Tier III, S4-ranked SGCN in LA.

Host Plants: Generally, yellow thistle and bull thistle are reported to be the larval food plants. Yellow thistle grows in open and disturbed places, flowering from Mar to June, often later farther south. Despite the name, in LA, the flowers are very often purple.

BRUSHFOOTED BUTTERFLIES (Nymphalidae)

American Snout (*Libytheana carinenta*)

ventral

dorsal

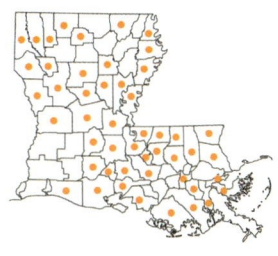

Description and Behavior: This unique butterfly is the only one in LA with extended labial palpi. When it lands, it will immediately fold its forewings under its hindwings and assume a very convincing leaf-like position with the extended palpi adding to the disguise as the leaf's stem. Its flight pattern is like a very large Harvester—fast, erratic, and hard to follow. It will often perch on dirt or shell roads, mostly at puddles but, at times, even when the roads are dry. I have occasionally found it taking nectar at Brazilian vervain, ligustrum, and Chinese privet. This butterfly has on numerous occasions landed not only on my person, but also on my net as if daring me to catch it.

My Records: Most of my records are west of the Mississippi River. I have found it regularly during the warm months at Indian Bayou WMA and Thistlethwaite WMA, where its numbers can fluctuate greatly. I have found it abundant at times, particularly after a drought followed by rain, with great numbers at puddles on the roads. At other times, only one or two are seen. It is multi-brooded with at least four broods, possible even a fifth if the weather remains warm into Nov. (One year I saw eight fairly fresh individuals in mid-Nov, at Cypremort Point SP.) It is not a regular inhabitant of the marshes and coastal prairies of CAM; my only record (three seen) was at Sabine NWR in July 2013. It also seems to be less common in the piney woods of central LA, where I once found several in Apr near one of the bayous in the NAT Kisatchie unit. These were on a dirt road or perched in the sun on hackberry trees growing by the bayou.

Distribution and Abundance: Although it has been recorded from most of the state during every month (with few reports in Dec and Jan), its primary flight season is late Mar to Sept. In CAD, there are records from all months, most common in Mar to May. In the Cajun Prairie region, large numbers were reported in May and lesser numbers in Apr, July–Oct, and Dec. Records from southeast LA include the New Orleans, Bonnet Carré, and Pearl River counts. Brou (1974) reported catching 167 in SJB, in ultraviolet light traps. Farther south, it occurs on an uncommon basis in LAFO from Mar through Dec.

Host Plant: In LA, as elsewhere within its range, it feeds on hackberry trees.

BRUSHFOOTED BUTTERFLIES (Nymphalidae)

Monarch (*Danaus plexippus*)

ventral

dorsal male

Description and Behavior: This famous butterfly is present in LA year-round, more common in the fall as it moves toward the Gulf of Mexico during migration. It gathers in large numbers along the coast in VRM and CAM in Oct, taking nectar from goldenrod and other wildflowers, fueling up before mak-

ing the trek across the Gulf of Mexico. During mild winters, Monarchs have remained all winter in locations such as Grand Isle and CAM, during which time they continue to reproduce. Gary Ross (2001b, 2010b) has published several excellent articles on the Monarch's use of oil rigs and platforms in the Gulf of Mexico as resting points while making the trip south.

My Records: While I regularly see this butterfly west of the Mississippi River (more than twenty-five at Freshwater Bayou Lock Rec Area on Oct 26, 2014), it is most regularly seen at Thistlethwaite WMA with its open fields, a steady supply of flowers blooming year-round, and a good supply of aquatic milkweed. I also find it regularly at various nurseries around the city of Lafayette, feeding on Mexican milkweed.

Distribution and Abundance: State records are from all months and suggest two annual peaks of abundance, Mar into Apr and then late Sept into early Nov. In CAD, it has been recorded from mid-Mar to the end of Nov, abundant in late Mar and early Apr, then again in Sept and Oct. There was one sighting during the Cajun Prairie survey for Jan, good numbers from Mar to May, small numbers from July to Sept, and then a big jump (eighty-six seen) in Oct, returning to small numbers again for Nov and Dec. It has been found in the Felicianas from Mar into early Dec, mostly uncommon, but locally common at flowers in fields during the migration from Mexico in late Mar and early Apr, and then again in Oct during the return trip. Israel (1981) felt it had only one brood in that region. It has been regularly recorded over the thirty-eight years of the Pearl River count (typically, June or July), with a high of twenty-five in 2006. In southernmost LA, records are mostly from Grand Isle and Galliano, including every month except July but primarily in the fall. It was listed in 2015 as a Tier III, S4-ranked SGCN in LA.

Cunningham verbally reported a rare white form, *nivosus*, in mid-Nov 2016, at Cocodrie. Believed to be caused by a recessive gene, this form has been reported on several occasions in FL. It is also regularly present in HI, where, on the island of Oahu, about 10 percent of Monarchs are white.

Host Plants: In LA, this butterfly feeds on Mexican milkweed, butterfly weed, and other milkweeds.

BRUSHFOOTED BUTTERFLIES (Nymphalidae)

Queen (*Danaus gilippus*)

ventral

dorsal male

Description and Behavior: The Queen resides across the extreme southern portions of the US. The southeastern subspecies (*D. g. berenice*) ranges from FL around the Gulf of Mexico to the MS Valley, and the southwestern subspecies (*D. g. strigosus*) ranges throughout TX into the Desert Southwest. *Strigosus* is paler

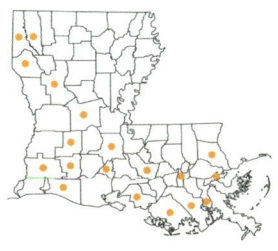

and has white scaling bordering the narrow black veins of the hindwing. Both subspecies have been reported in LA. Queens are common in some years, rare in others, primarily arriving along the Gulf Coast in late summer and fall. I address the Mullerian relationship between this species, the Monarch, and the Viceroy in the Viceroy's species account.

My Records: There was a breeding colony at Thistlethwaite WMA with records over several years. In mid-Aug 2010, I recorded thirteen Queens there, and over three additional weekends in Aug and Sept, I returned and counted sixteen, thirteen, and seventeen Queens, mostly in the same area. The colony at Thistlethwaite WMA in 2010 was the *strigosus* subspecies. In Oct and Nov 2012, I saw several in west CAM, both at Sabine NWR and Peveto Woods. In late June 2013, I saw a single male at Lacassine Pool. I also saw one at Abita Creek Flatwoods Preserve (in STA) in the pitcher-plant bog section in Aug of the same year. I did not see another until Nov of 2016 in lower SMA.

Distribution and Abundance: While early records suggested this danaid was rare in LA, the Season Summaries of the Lepidopterists' Society and NABA count results over the last fifteen years reveal multiple sightings, mostly in the coastal areas, including an established colony on Grand Isle. In north LA, it is a stray with sightings in Sept, Oct, Nov, and even into mid-Dec. In the central LA region, specimens were recorded at Kisatchie NF in RAP in Aug and Sept. In the Cajun Prairie region, there have been reports in July, Aug, and Oct at locations such as Vidrine's Cajun Prairie Gardens and the Eunice Cajun Prairie. It has been reported as part of the Cameron counts in 1990, 1998, and 2000. It is uncommon in the FL Parishes, with records from Gary Ross's Baton Rouge garden and Honey Island Swamp WMA (Sept). In the River Parishes, it was sighted in Edgard (late Oct). It was recorded on the New Orleans count in July 1993. Farther south, sightings have been common in JEF (Jan and Nov), LAFO (Jan), and TER (Sept at Timbalier Island). For example, a total of sixty-four were recorded between 1988 and 2013, primarily from Grand Isle and Golden Meadow, May through Jan, with most sightings between Aug and Nov.

Host Plants: In LA, like its cousin, the Monarch, it uses various species of milkweeds. At Thistlethwaite WMA, Queens were reproducing on aquatic milkweed, present in the roadside ditches and power-line cuts.

BRUSHFOOTED BUTTERFLIES (Nymphalidae)

Gulf Fritillary (*Agraulis vanillae*)

ventral

dorsal male

dorsal "blonde" male

Description and Behavior: The Gulf Fritillary is a tropical species that inhabits portions of the southeast US. Not a true "fritillary," it was so named because of the silver spots on its ventral hindwing, but it is actually a member of the Longwing subfamily. Longwings have a longer life than most, and this one can live from four to six weeks. Also, Ross (2001a, 2005b, 2010c) has opined that Longwings, including this species, exhibit an ability to learn and remember physical features. Typically, males are bright red-orange with black markings. Females are slightly browner with heavier black markings. Some females can be quite dark. The larvae are aposematic, protected by plant toxins taken from the host plant. Williams (1990) asserted that the adults are also one of several butterflies possessing "the stereotypical orange-and-black pattern that advertises unpalatability in nature." Gulf Fritillaries can often be found flying with Viceroys, and in some locations (Thistlethwaite WMA) with Queens as well. The mimicry relationship between Queens and Viceroys is well documented as Mullerian in nature. Do the occasional darker female Gulf Fritillaries gain protection through mimicry of one or both of the other two? If, as some suggest, the Gulf Fritillary is also distasteful, then might all three benefit from a triangular Mullerian relationship?

Following hard, freezing winters, this butterfly is noticeably absent early the next spring, only to slowly begin to reappear during the summer. After

BRUSHFOOTED BUTTERFLIES (Nymphalidae)

Gulf Fritillary (*Agraulis vanillae*)

(continued)

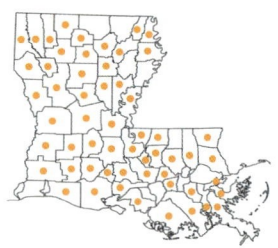

a mild winter, I have seen it as early as Feb. Ross (2013c) has suggested that the reduction in numbers each winter was more a result of starvation and dehydration, and that, in fact, some of these fritillaries across south LA do survive the state's bouts of freezing weather. At times, particularly in late fall, this is the most common larger butterfly flying in LA. During the Cajun Prairie survey, over five hundred were seen in the month of Aug and over six hundred in Sept. This butterfly loves open fields, disturbed areas, and city gardens where nectar sources are available. The males patrol constantly. Both sexes readily come to many kinds of flowers.

My Records: I have seen it in every month of the year, extending well into central LA and farther north. In late June 2009, in the city of Lafayette, I saw an oddly colored Gulf Fritillary, feeding at lantana. As the photo herein reflects, the butterfly was not an albino, but more of a "blonde."

Distribution and Abundance: In north LA, it has been reported from May until Nov, abundant Aug through Nov. It was the most common butterfly during the Cajun Prairie survey, with over sixteen hundred seen, flying in every month but Feb. At Asphodel Plantation, it is present from Jan to Dec, during which, at times, it is abundant, most common from Sept to Nov. Strangely, Israel (1981) found it to be rare to uncommon during his thesis survey of Tunica Hills, with a sighting in mid-July, then from early Oct to late Nov. In southernmost LA, there have been multiple sightings in LAFO and JEF during every month but Apr, with several sightings from Dec through Feb.

Host Plant: The larval food plants are both native and ornamental passionflowers.

Zebra Heliconian *(Heliconius charithonia)*

ventral

Description and Behavior: This unique member of the Longwing subfamily is a vagrant to the Houston area, appearing in late summer and establishing breeding colonies by using cultivated passion vine. To the east, breeding colonies have been reported along the MS and AL Gulf Coast. Ross (2010c) has documented the learning abilities of this butterfly. Specifically, he described how several individuals of this species, reared by him, flew a daily, repeated route around his home and garden as they moved in search of nectar sources. Ross (2013c) felt LA's winters were "just a bit too severe to sustain viable long-term colonies." He further indicated that he felt other sightings around the state could have been the result of occasional colonization or butterfly releases.

My Records: I saw two in LAFA on consecutive weekends in late Nov and early Dec 2005, and now wonder if they were present through human aid, such as a wedding release.

BRUSHFOOTED BUTTERFLIES (Nymphalidae)

Zebra Heliconian (*Heliconius charithonia*) *(continued)*

dorsal

Distribution and Abundance: There are three very old LA records: ORL (1863), JEF (1894), and TER (1919). More recent records are also from the southeastern section of the state. In the FL Parishes, it has been seen rarely in the Asphodel Plantation and Jackson areas from Oct to Dec. Other sightings in that region include the Audubon Commemorative Area near St. Francisville (Oct) and near Abita Springs. In the NO area, Brou, verbally and then supplemented by e-mail, reported collecting fifty-one specimens that he identified as the FL subspecies, *H. c. tuckeri*, taken in late July to early Aug of 1995 and 1996 from a breeding colony along the Westbank Expressway at Marrero (JEF). The LSU State Collection contains a specimen from the Prairieville area in Nov 1995. Farther south, specimens were seen in and around Houma in Sept and Oct of 1997; however, a little more than a year prior to those sightings there had been a release associated with a local butterfly exhibit. Trahan suspected the same of his sighting in CLA in 2002. In Oct 2014, there was a sighting back in the NO area that was also suspected not to be a natural occurrence. In the southwestern portion of the state, McMillian by e-mail reported this species at his nursery in BEAU from Oct to Dec 2016. In Nov 2016, a single individual was sighted (with a photo to substantiate) at Peveto Woods in western CAM. These sightings suggest a migration from south TX.

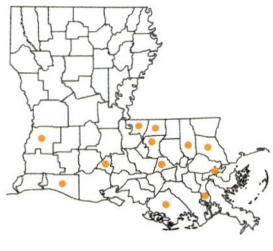

Host Plant: The larval food plant is primarily passion vine, particularly common purple passion vine. Brou reported that it used yellow passion vine in JEF. Specifically, by e-mail he advised that he had found the white larvae of this longwing on that plant but not on common purple passion vine, which was much more common in the area and had Gulf Fritillary caterpillars on it.

BRUSHFOOTED BUTTERFLIES (Nymphalidae)

Variegated Fritillary (*Euptoieta claudia*)

ventral

dorsal

ventral (winter form)

Description and Behavior: This butterfly is the sole LA member of its genus. Cech and Tudor (2005) indicated that, while the Gulf Fritillary falls on the Heliconian side of the line between fritillaries and Heliconians, the Variegated Fritillary falls on the fritillary side of that line. This butterfly is never bright orange, but is a dullish yellow-orange. It looks like a greater fritillary dorsally and a lesser fritillary ventrally. Early in the year, this butterfly typically is smaller and darker than a Gulf Fritillary, but by late summer it is about the same size. It prefers open, disturbed areas, typically flying low to the ground, just above the level of vegetation.

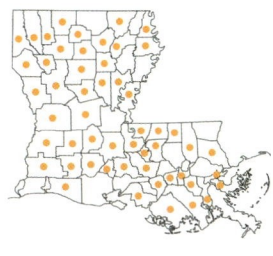

My Records: I have found it all over the state in every month but Jan and Feb. Its occurrence can be erratic. One May, at Myrtles Plantation in St. Francisville, it was extremely abundant, but then noticeably absent the next time I was there. Good places to look for it include the levee road in Indian Bayou WMA or the internal roads at Thistlethwaite WMA. I have also found it in the NAT unit of Kisatchie NF from Mar and Apr and then again from Sept into Nov.

Distribution and Abundance: Records throughout the state show this butterfly to be common, more abundant in south LA. CAD records include all twelve months of the year, more common in Apr and May. Other north LA locations include the Bayou Pierre Unit (Apr and Oct), Kisatchie NF in WIN (late Sept), Lake Claiborne SP (Apr), Sparta Quad (July), Bodcau WMA, and Cane's Landing (Mar, May, and Aug through Sept). In the Cajun Prairie region, it was reported in seven parishes as the fourth most common butterfly, from Apr to Dec with very high numbers from June to Sept. In the FL Parishes, it has been recorded at Grand Bay in the Baton Rouge area and near Erwinville. One Dec, it was seen at Greenwood Park. There are four specimens in the LSU State Collection, taken on January 1, from the St. Gabriel area. It was uncommon during the Tunica Hills survey, with three broods from early June to mid-July, then mid-Aug into early Dec. At Asphodel Plantation, it flies from Mar through Nov, most common in May and Sept into Nov. It has also been recorded eighteen times during the Pearl River annual count, with eighty-one observed in 2006. Farther south, there are records for Grand Isle, Galliano, and Larose from Mar to July and Sept to Oct.

Host Plants: In LA, the host plants are passion vine, violets, and flax.

BRUSHFOOTED BUTTERFLIES (Nymphalidae)

Diana (*Speyeria diana*)

ventral female

dorsal female

dorsal male

Description and Behavior: This large fritillary is very dimorphic. The female is large, and is reported to be one of several butterflies that mimic the Pipevine Swallowtail. This mimicry, from the dorsal aspect, can be clearly seen in the photos of the two insects. Ventrally, the female Diana lacks the bright orange-red spots that are so distinctive with the Pipevine. Ross (2008, 2011) has indicated that he suspects her dark color may, in fact, serve two purposes. While her similar coloring to the Pipevine may provide her with a degree of protection, that coloration could also serve a thermal regulation purpose, allowing

her to absorb solar heat in late fall as she begins to lay eggs. The male is smaller than the female, about the same size as a Great Spangled Fritillary, the only other butterfly with which a male Diana might be confused. In flight, it can be difficult to tell them apart, but the male Diana's orange is brighter. Of course, the male Diana has no ventral silver spots.

Females, more so than males, frequent the shade, flying about eight to ten feet off the ground in open woods. The females will land on leaves in sunny spots, absorbing the sun's heat with her wings open. Her flight is direct but leisurely unless disturbed, at which time she will quickly disappear into the trees. Males fly a route most of the day. They will often be seen out in open areas, crisscrossing from one wooded area to the next. They will pause to take nectar, but not nearly for the length of time that a female will stop. In the late afternoon, both sexes can be found feeding, side by side, at purple coneflowers or butterfly weed.

My Records: I have seen this unique butterfly numerous times at Rick Evans/ Grandview Prairie WMA in Hempstead County and Stone Road Glade Natural Area in Howard County (both in AR), but not in LA. Those viable populations in Hempstead County (including Nacatoch Ravines Natural Area) are located less than one hour from the AR-LA border, so it is not inconceivable that this butterfly could turn up in LA again.

Distribution and Abundance: Mather and Mather (1958) reported a 1953 sight record of this fritillary at Tallulah in MAD, across the Mississippi River from Vicksburg, which they considered to be reliable. I am unaware of any further sightings from either state in the interim.

Hershel Raney's website "Butterflies of Arkansas" represents it as present in Calhoun and Bradley counties in south-central AR.

Host Plant: Violets.

BRUSHFOOTED BUTTERFLIES (Nymphalidae)

Bordered Patch (*Chlosyne lacinia*)

ventral female

dorsal

CRESCENTS AND CHECKERSPOTS (Tribe melitaeini)

Description and Behavior: This south TX/western US butterfly is known to wander north out of south TX, becoming common in the Victoria area and a regular stray in the Dallas, Harris County, and Freeport areas of TX. In peak years it can reach into OK and even KA. For example, it has been reported at Red Slough WMA in extreme southeast OK. This  butterfly flies low to the ground, acting much like the Silvery Checkerspot or Texan Crescent. It is a colonial butterfly, which, if breeding, can be abundant. Males readily perch, and both sexes take nectar from numerous types of low-growing flowers. Known to be variable in appearance, including specimens that are predominantly black, it might be confused dorsally with a Texan Crescent, but the two are quite different ventrally.

My Records: I have found this pretty little butterfly in TX, as far north as San Antonio, and as far east as the Corpus Christi region, but not in LA.

Distribution and Abundance: As I was preparing this book, I had listed this butterfly as one that might ultimately be found here, so I was not surprised when it was reported via posts to the LA listserv along LA's western border in CAD by Jeff Trahan and Rosemary Seidler at Eddie Jones Park in mid-Oct 2012.

Host Plant: The most common larval food plant identified is sunflowers, which grow both in the wild and commercially in much of LA.

BRUSHFOOTED BUTTERFLIES (Nymphalidae)

Gorgone Checkerspot (*Chlosyne gorgone*)

ventral

dorsal

CRESCENTS AND CHECKERSPOTS (Tribe melitaeini)

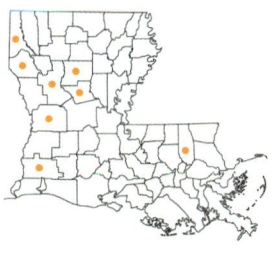

Description and Behavior: Similar in size and dorsal appearance to the Pearl Crescent, this checkerspot is quite distinctive ventrally with a post-median row of black dots encircled in orange and a zigzagged pattern of lines inside those spots. The males are smaller than the females. The eastern subspecies, *C. g. gorgone*, is rare and extends west to AL (with no records from that state in over fifty years). The western subspecies, *C. g. carlota*, is more common, extending into TX and AR and reported to be common to uncommon in the piney woods region of east TX along the LA border. I have not found any report of which subspecies has been found in LA. The western subspecies is multi-brooded.

My Records: I have searched the central LA region without success. My closest record is Stone Road Glade Natural Area in Howard County, AR.

Distribution and Abundance: Many past reports on this butterfly are erroneous or suspect. Lambremont's reports have subsequently been identified as *P. phaon*, not *gorgone*. Strickland (1972) reported that he found it in central LA and the Tunica Hills region, but his field records clearly confuse and/or combine *gorgone* and *phaon*. During the Cajun Prairie survey, this checkerspot, identified as the Great Plains Checkerspot, was recorded from multiple stations from Mar until July, with Apr and July being the peak months (nineteen in Apr). That survey had no records whatsoever for either Silvery Checkerspots or Phaon Crescents, both of which are regulars in this region, causing me to wonder if this species had been confused with *P. phaon* and/or *C. nycteis* (Allen and Vidrine 1990). I learned that only one photo existed of this species from the survey, a non-diagnostic dorsal shot, and the possibility of misidentification was real. While possibly some of the butterflies recorded were, in fact, *gorgone*, in all probability the great majority were not. The survey participants have verbally advised that this butterfly has not been seen in that region in recent years.

The first confirmed records were several specimens from CAD and DES in 1958. It was also reported as abundant in GRA, NAT, and WIN in 1962 and 1964. Several of those specimens are within the LSU State Collection. During the early 1970s, it was found in the NAT unit of Kisatchie NF. Brou verbally reported finding this checkerspot in the FL Parishes, the only record from that section of the state of which I am aware. He had thirty-one specimens in his

Gorgone Checkerspot *(Chlosyne gorgone)* *(continued)*

collection, both males and females, taken during the first week of Apr 1979, near Fluker on private property no longer accessible to the public. I found no further reports of this butterfly within the state until late May 2013, when one was photographed at Cooter's Bog as part of the 2013 Fort Polk count. The photographs were submitted to and accepted by BAMONA as authentic.

Host Plants: The western populations use a variety of plants in the sunflower family as larval food plants. Other plants used include the black-eyed Susan and ragweed families. The eastern population has been reported to use woodland sunflower from the Asteraceae family. In GA, it has been found on giant or tall sunflower.

Silvery Checkerspot (*Chlosyne nycteis*)

ventral

dorsal

BRUSHFOOTED BUTTERFLIES (Nymphalidae)

Silvery Checkerspot (*Chlosyne nycteis*) (*continued*)

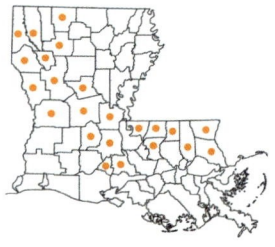

Description and Behavior: The Silvery is highly colonial, and can be locally abundant. Ross (2001c) conducted an excellent study on this checkerspot, describing it as very local in LA, usually associated with its recorded host plant, wingstem crownbeard, which grows in disturbed habitats. It closely resembles the Pearl Crescent but is larger (particularly the females) and more darkly marked dorsally. There is also a round spot with a silver or white center on the dorsal margin of the hindwing that is not present on the Pearl. Unless disturbed, the flight is slow and leisurely. The males patrol less than a foot over the ground cover in a flap-and-glide fashion. The females are not seen as often except when taking nectar from low-growing flowers like frog fruit. When at a flower, they are easily approached.

My Records: My records are from Mar through Oct with flights that are sporadic rather than constant. At times there are hundreds on the wing at Thistlethwaite WMA, with some of the largest females I've ever encountered. There is a healthy colony at the LA Arboretum. Two adults and several viable caterpillars were recorded during the 2001 Indian Bayou WMA count, and I have seen the Silvery there at other times but not in several years.

Distribution and Abundance: LA records are generally distributed throughout the state, Mar into Oct. In CAD it flies from late Mar to early May, at times common. Other northern records are from Haynesville, at Wallace Lake (Mar), at Bodcau WMA (May), and Sparta Quad (July). There are a few records from central LA and from the Acadiana region, including a large colony (eighty-five seen one day in early July) at Acadiana Nature Park, with records from late June through mid-Nov. It flies at Asphodel Plantation from Mar to Oct, most common from July to Sept, but it is generally uncommon in the Tunica Hills area over two broods from early Apr to early Oct. Other FL Parishes records include Greensburg, Fluker, and Honey Island Swamp WMA (Sept). It was also recorded during the June 2011 Pearl River count.

Host Plants: This checkerspot uses various herbaceous members of the aster or composite family (Asteraceae), including wingstem crownbeard, great ragweed, and purple coneflower (Ross 2001c). The larval food plants reported to be used in southeast TX include sunflowers and black-eyed Susans.

Phaon Crescent (*Phyciodes phaon*)

ventral male

dorsal male

Phaon Crescent (*Phyciodes phaon*) (*continued*)

dorsal female (top) and male (bottom)

Description and Behavior: This pretty little butterfly is, at times, very common. Occasionally seen as early as Apr, it increases in number during the summer until, by late fall, it can be abundant in locations such as CAM. It prefers damp, open areas where frog fruit grows. In some of the older guides it is called the Map Butterfly based on its ventral hindwing pattern. It can be confused with the Pearl Crescent, with which it often flies. Both fly in a flap-and-glide manner, low to the ground. The males patrol, occasionally stopping at flowers. The females are seen more often at flowers and for longer periods of time. While at a flower or perched, the Phaon can be identified by its darker edges and more reddish orange dorsal coloring.

My Records: The first time I found this butterfly was in early Nov at Sherburne WMA, and it is present both there and at Indian Bayou throughout its long flight season. It is always present at Thistlethwaite WMA and Peveto Woods in

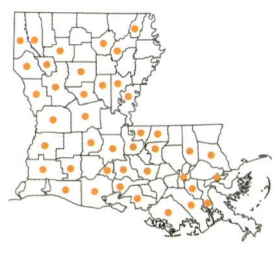

good numbers. In fact, in Oct it might be the most numerous butterfly on the wing in CAM and SLA, two parishes with radically different habitats but lots of frog fruit. It can be common at the Burbank soccer fields in EBR, also in early Nov. Along the Mississippi River corridor, I found a healthy colony in the ditch along the main road at Bayou Cocodrie NWR in CON.

Distribution and Abundance: This butterfly was initially reported in LA as *P. gorgone*, so older records may be unreliable. In CAD, it has been recorded from all months except Feb, common to abundant from Apr into Oct. This crescent was not reported during the Cajun Prairie survey, even though it is very common in SLA, and has been found at Bunchs Creek in ALL. In the Felicianas, it was rare (only in Apr) at Asphodel Plantation, but common, even abundant at times, in WFE, from late Mar to late May and late Aug to mid-Nov. Over the course of thirty-eight years, it has been reported only three times (each time in small numbers) during the Pearl River count. It has also turned up twice (again, in small numbers) on the Catahoula count. In southernmost LA, records include Grand Isle, Golden Meadow, Port Fourchon, Galliano, and Larose, flying from Jan into Nov, seen more often in the late fall.

Host Plants: In LA, it uses members of the frog-fruit family, also known as carpetweed.

BRUSHFOOTED BUTTERFLIES (Nymphalidae)

Pearl Crescent (*Phyciodes tharos*)

ventral female (top)
and male (bottom)

dorsal male

Description and Behavior: This is one of the three most common butterflies in LA (the Common Buckeye and the Carolina Satyr being the other two). A colonial butterfly, it can be found virtually anywhere—wooded areas, open fields, coastal areas, marshes, swamps, piney hills, and city parks. It is primarily orange in flight, flying in a flap-and-glide manner low to the ground. It will perch with its wings held horizontally. The males are smaller and constantly patrol for females. The females can be twice as large as the males and are most often found on low flowers like frog fruit, clover, asters, and dandelions. There are two color phases that appear to depend on the season. In the spring and fall, it is darker orange and brown ventrally, while in the summer it is much more yellow.

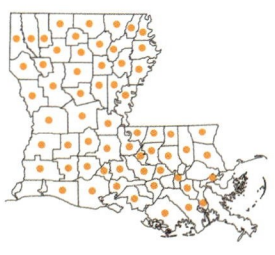

My Records: I have not yet seen one in Dec. My latest sighting was the last day of Nov, at Abita Creek Flatwoods Preserve.

Distribution and Abundance: This butterfly can be found throughout the state, in all months except the coldest days. In CAD, there was a single sighting in Jan; it has been found to be common from Mar to Dec. It was seen in good numbers during the Cajun Prairie survey in five parishes from Mar to Nov, with in excess of one hundred sightings from Sept to Nov. There were only five sightings in Dec. During Israel's (1981) survey of the Tunica Hills region, it was described as the "most ubiquitous species in the hills." Found at multiple low-growing flowers (twenty-seven different species recorded) and mud, there were records from late Jan to early Dec over four broods in that region. Farther south, records are from JEF and LAFO from Jan 29 into mid-Nov.

Host Plants: In LA, it uses members of the sunflower and aster families. It has also been reported as found on Eupatorium species.

BRUSHFOOTED BUTTERFLIES (Nymphalidae)

Texan Crescent (*Anthanassa texana*)

ventral "Seminole"

dorsal "Seminole"

dorsal "Texana" male

Description and Behavior: This is a southern species with two subspecies, *A. t. texana* in the US Southwest and *A. t. seminole* in the US Southeast. Both have been found in LA. Some sources consider the Seminole Crescent to be a separate species, *A. seminole*, based upon habitat preference and other distinguishing characteristics; however, the majority include *seminole* as a subspecies of *A. tex-*

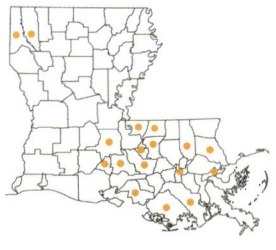

ana, most common in the FL Panhandle and central FL. That eastern subspecies was listed in 2015 as a Tier III, S3-ranked SGCN in LA. It is multivoltine, with three to four, possible five, annual broods. The adults fly from Apr to Nov. Its preferred habitat includes trails through moist, deciduous woods and shaded but open, disturbed areas near the same kinds of woods. It is often found immediately adjacent to water. It stays low to the ground, perching on low-growing grasses and weeds.

My Records: I have seen it irregularly. My first records were from a colony in LAFA, now gone; however, another colony exists in that parish at Acadiana Nature Park (Mar to Nov). I have found it sporadically at Indian Bayou WMA and Thistlethwaite WMA, but not recently. I found two small colonies at Bayou Teche NWR in July (the North Bend Unit and the Palmetto Trail parking area). I found two in early Nov at boneset in a power-line cut at Mary Ann Brown Preserve. I also found a couple at Weyanoke in early May. All were the Seminole subspecies.

Distribution and Abundance: First reported in LA in 1958 (CAD), that specimen was the western subspecies. All other reports are of the eastern subspecies. Ross (2005a) has reported on a recurring colony in and around Baton Rouge, including at Bluebonnet Swamp Nature Center. For a detailed study of this butterfly's life history, I would encourage the reader to review that article. Other reports are primarily from the southeastern sections of the state, with records from Asphodel Plantation (May and Oct), Weyanoke (Oct), along the Manchac Road (Nov), near Erwinville (Aug), and Greenwood Park (June, Aug–Sept, Oct [with eight seen], and Nov). In the FL Parishes, there are records from Abita Springs, with dates in Mar, May, Oct, and Nov. It has been reported at Edgard in SJB. Pearl River count results from 1993 through 2000 indicated the presence of a colony at Pearl River WMA. It was also reported once on the 1992 New Orleans count. Farther south, a healthy colony was recorded near Houma, and it has been reported in the Larose area, flying in Feb, June, and Sept into Nov.

Host Plants: Ross (2005a) determined its preferred larval host plant to be the lance-leaved waterwillow, but it will also use several other members of the acanthus family such as Brazilian plume, shrimp plant, branching foldwing, and King's crown.

BRUSHFOOTED BUTTERFLIES (Nymphalidae)

Common Buckeye (*Junonia coenia*)

ventral

dorsal

ventral
[rosa]

Description and Behavior: This is the second of three that I consider to be LA's most common butterflies. Distinctive for its multiple dorsal eyespots, it is primarily found in open fields and secondary-growth areas but can also be found in city parks, yards, gardens, soccer fields, and even gravel or shell parking lots. The males are territorial, aggressively defending small patches of ground from

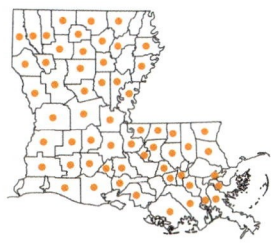

other males as well as other intruders, even dragonflies. The males primarily perch on the ground,
but will also use grass. Both sexes have a flat, flap-
and-glide flight, no more than a couple of feet off
the ground. During the Tunica Hills survey (1981),
Buckeyes were recorded at twelve different species
of flowers. The Common Buckeye is multi-brooded,
with probably four broods per season. Ventrally, the summer broods are light,
tawny colored with small spots. In the fall, the ventral hindwing of the rosa
form is a beautiful shade of maroon.

My Records: I have found this butterfly all over the state. It has been recorded at
every NABA count conducted at Thistlethwaite WMA and Indian Bayou WMA.
During the 2008 Kisatchie/NAT count (late May), in an area of open pine flats,
numerous Buckeye caterpillars were encountered in the grass and also on a
thick, low-growing vine so prevalent that walking was difficult. The caterpillar
numbers significantly exceeded the number of adults. Also at the Kisatchie unit
in NAT, in early Mar 2012, I found a meadow full of small dark butterflies that
I did not initially recognize. They were unusually small Buckeyes, more than
fifty. I suspect the size of this spring brood had been influenced by an extreme
drought in the region during the summer and fall of 2011. The latest I have seen
this butterfly was on the last day of Nov, at Abita Creek Flatwoods Preserve.

Distribution and Abundance: Buckeyes can be abundant during much of the
year. In north LA (CAD), this species has been recorded in every month. It
was the second most common butterfly found during the Cajun Prairie survey
with a total of 927, again seen in every month. At Asphodel Plantation, records
were for every month except Feb, most commonly from Aug to Nov. During
the Tunica Hills survey, four broods were noted from late Mar into early Dec.
It even showed up five times in Brou's light traps in SJB. Farther south, it has
been reported during every month except Feb at Grand Isle as well as in multiple locations in LAFO.

Host Plants: In LA, it feeds on snapdragons, snapdragon vines, and toadflax. In
the Felicianas, the conspicuous larvae were found in Sept and Oct at pink glove.
In southeast TX, larval food plants are reported to include Gerardia, plantain,
false foxglove, and ruellias.

BRUSHFOOTED BUTTERFLIES (Nymphalidae)

White Peacock (*Anartia jatrophae*)

ventral

dorsal

Description and Behavior: This is a tropical butterfly that has made its way into LA on a few occasions. Ross and Welden (2003) suggested that strays are occasionally blown into the state by tropical storms and establish breeding colonies that survive for a period of time. It acts much like the Common Buckeye, with a flight close to the ground in a flap-and-glide fashion. Males perch on patches of ground or stalks of grass, darting out at anything that passes. Its preferred habitat is mostly disturbed areas, out in the open. When present, it can be very plentiful.

My Records: I've seen this butterfly in FL as far north as Jacksonville as well as in the Rio Grande Valley, but not yet in LA.

Distribution and Abundance: LA records indicate this species periodically migrates into LA from the east. The first reports, by Brou et al. (2008), from mid-Nov into Dec 1999, were of the FL subspecies, *A. j. guantanamo*, at Golden Meadow, with fifty seen in one day. The next year, more were seen at Grand Isle in late Aug, and then in LAFO in Oct. In 2005, one was found in Oct and another in Nov, back at Golden Meadow. During the following Oct, several were seen near Grand Isle SP, continuing into mid-Nov. This colony was found in the proximity of roundleaf bacopa, a low ground cover that produces small white flowers, reported as the larval food plant. During Sept and Oct 2013, multiple specimens were reported at Galliano and at Bayou Sauvage NWR in NO (thirteen on October 6). In mid-Nov 2013, seventy were reported at the latter location, in the area of the marsh boardwalk, where roundleaf bacopa was growing extensively. Other locations during 2013 included Diamond (late Oct) and below Venice (mid- Nov). Despite near-freezing weather during the last week of Nov 2013, two were still flying at Bayou Sauvage NWR, and ten or more were reported during the Grand Isle Christmas bird count in late Dec 2013.

Host Plants: In two separate locations, roundleaf bacopa was identified as the larval food plant. In the South in general, members of the frog-fruit family are used as well as water hyssop, snapdragons, and ruellias.

BRUSHFOOTED BUTTERFLIES (Nymphalidae)

Question Mark (*Polygonia interrogationis*)

ventral

ventral
(summer)

dorsal (fall)

dorsal (summer)

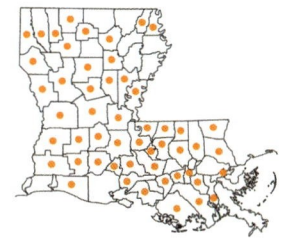

Description and Behavior: This is primarily a woodlands butterfly, found in the forest's interior and along its edges, occasionally venturing out into adjacent open areas. It is a nervous butterfly, quick to fly and fast on the wing. It is also curious and will readily leave its perch to investigate passersby. At times, males maintain territories from favorite perches. It is not uncommon to find them at puddles, damp spots, mammal scat, and roadkill on roads and trails. There are two seasonal forms: spring and fall with orange dorsal hindwings containing dark and light spots, and summer with almost entirely dark dorsal hindwings that reflect a purple sheen when freshly enclosed. Ventrally, the fall brood is a lighter, mottled brown while the summer brood is a flat, grayish brown like an old, dead leaf. On the fall brood, the tails are longer and more pointed. The fall brood overwinters and can occasionally be seen during the winter. The females are typically larger than the males, with more squared and less angular forewings.

My Records: My records are throughout the state, usually in small numbers. I have not yet seen it in Dec, Jan, or Feb, but I am rarely in the field then. On occasion, usually during wet springs, both adults and caterpillars of this butterfly have been found in large numbers on the roads, in the trees, and even in the grass. I have experienced these "irruptions" at Thistlethwaite WMA, Indian Bayou WMA, and Bayou Teche NWR.

Distribution and Abundance: In CAD, it is mostly common, with a single sighting in Jan, then regular sightings from late Feb into early Dec. It was quite common in the Cajun Prairie during the survey of that region, with 186 recorded from Feb through Nov and peak numbers in Apr and May. Along the Mississippi River, records include Gonzales, near Erwinville, and at Greenwood Park (Dec). Records in the Felicianas are from late Jan to mid-Dec over four broods, where it can be common along open trails. Brou has taken large numbers in his traps in SJB and STA, even during the winter months. He has opined that this anglewing has five broods in LA. It is a regular at the Pearl River count, with a high of 32 in 2001. In south LA, specimens have been recorded from mid-Feb into early Oct.

Host Plants: In LA, as elsewhere, it feeds on nettles, hackberries, and elms such as American elm. In southeast TX and south FL, larval food plants are reported to include false nettle and hops.

BRUSHFOOTED BUTTERFLIES (Nymphalidae)

Eastern Comma (*Polygonia comma*)

ventral

dorsal (fall form)

dorsal (summer form)

Description and Behavior: This anglewing is similar in appearance to the Question Mark but is smaller and has a stubbier tail. Also, it has a "comma" shaped silver mark on the lower ventral wing with no "dot" as is found on the Question Mark. It has the same habits and frequents the same habitat as the Question Mark. It also has the same seasonal forms within the state as the Question Mark. It is never common here and has an erratic distribution.

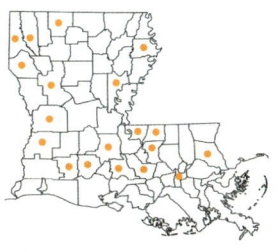

My Records: I often go several years without seeing one. I see it most often at Indian Bayou WMA during the summer. It can occasionally be found there after a rain, puddling on the road along the levee. I have seen singles in the NAT and RAP units of Kisatchie NF in early Mar that were fresh spring/fall forms. Two were seen at Thistlethwaite WMA during the 2013 count in mid-May.

Distribution and Abundance: The first reported LA specimen was in the FL Parishes (Mar). It is occasionally found in CAD from late Feb through the end of Mar with one additional sighting in May and another in Aug. North LA locations include Haynesville (feeding on rotting pears), Eddie Jones Park, Bodcau WMA (Aug), and Talla Bena. In the central LA region, it has been recorded at Kisatchie NF in NAT (Apr) and Sicily Island Hills WMA. Twelve were recorded in the months of Apr, June, and Sept during the Cajun Prairie survey. In Acadiana, aside from my records, there is a mid-Mar report from Acadiana Nature Park. It is mostly uncommon in the Felicianas, with records of four broods from mid-Feb into early Dec. Brou (2015) captured as many as fifty specimens over the years at traps in Abita Springs and Edgard, with records from Mar, Apr, June, July, Aug, Sept, Oct, and Nov, but most commonly in Mar, Oct, and Nov. He has proposed the existence of three broods in the FL Parishes—Mar to Apr, late June to early Aug, and Oct to Nov.

Host Plants: As throughout much of its range, in LA it uses elms, nettles, and hornbeam trees.

BRUSHFOOTED BUTTERFLIES (Nymphalidae)

Mourning Cloak (*Nymphalis antiopa*)

ventral

dorsal

Description and Behavior: This is another woodlands butterfly, typically seen along tree lines or coursing along roads within the forest. It will often stop and bask in sunny spots. Although it was reported to visit Chickasaw plum during Israel's Tunica Hills survey, I have not personally seen it at flowers. I have seen it several times at wet spots in the road or trail. It is not common in LA, but overall records suggest it is an uncommon resident in some areas.

My Records: My records are mostly singles in the spring, some tattered, suggesting they had hibernated through the winter. I have seen it only once in north LA (Eddie Jones Park, Apr). I have seen it on a recurring basis at Thistlethwaite WMA (Apr–May). In mid-Mar 2012, I saw five at Sicily Island Hills WMA (all of which were very tattered) and then a fresh male at Avery Island in mid-Apr. In late Apr of that same year, I saw four at Thistlethwaite WMA. Two days later, I saw eight at Tickfaw SP, the most I've seen in LA at one time. That same week I saw one (fresh) at the Catahoula Butterfly Garden. Additionally, three were reported during the Fort Polk count during this same time frame. The summer passed without sightings, but in Oct, I briefly saw one north of Freshwater Bayou Lock Rec Area. Also in Oct, another was reported in TER. Finally, a last one was found at Peveto Woods in mid-Nov. These sightings made 2012 an exceptional year for this butterfly. I also saw five in mid-Mar 2013 at Kisatchie NF in NAT.

Distribution and Abundance: Older records, primarily in the fall, suggested it migrated into LA from the North during that season; however, more recent records indicate it is established within the state. There have been several sightings in CAD in Feb, Mar, July, Aug, and Sept, and three sightings in RDR between Mar and May. In the central LA region, records include NAT (Kisatchie NF, Mar), GRA (mid-Feb), and CAT (mid-May). In the southwest region, there is a record from BEAU (Nov). In the Cajun Prairie area of Acadiana, it was reported at Chicot SP (early Mar) and at Vidrine's Cajun Prairie Gardens in ACA (late Apr). Another was seen in the Washington area (June). There are also several records from southeast LA. It was reported as rare at Asphodel Plantation from Feb to June and then Aug to Oct, most common in May. In WFE, records suggest three broods, late Feb to Mar, May, and Aug to Oct, most prevalent in the spring, but still uncommon. It has been recorded in SJB (May) and during the 2003 Global Wildlife count (Aug). It was also reported in the Galliano area in early Mar.

BRUSHFOOTED BUTTERFLIES (Nymphalidae)

Mourning Cloak (*Nymphalis antiopa*)

(continued)

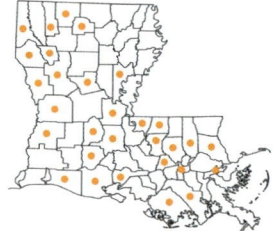

Host Plants: As in the rest of its range, it uses willows, ash, alder, basswood, birches, white poplar, elms, hackberry, maple, red mulberry, red chokeberry, cottonwoods, and Eastern hop-hornbeam trees.

Milbert's Tortoiseshell *(Nymphalis milberti)*

ventral

dorsal

Distribution and Abundance: This butterfly is typically found in the western US or much farther north than LA. It was found here only once, in Oct 1969, by Brou (2008b) at Edgard. It was netted a few hundred yards from the Mississippi River. Brou noted the closest records were in Missouri. As it would be unexpected here in LA, Strickland (1972) believed it had arrived here via river traffic.

BRUSHFOOTED BUTTERFLIES (Nymphalidae)

Red Admiral (*Vanessa atalanta*)

ventral

dorsal

Description and Behavior: One of our speediest brushfoots, this is a colorful, pugnacious butterfly. The red on its dorsal wings is distinctive, but its ventral side is equally colorful, with red, pink, and blue hues mixed in with the grays, blacks, and browns. It is multi-brooded and adaptable to numerous habitats. It is not actually an admiral; its closest relatives

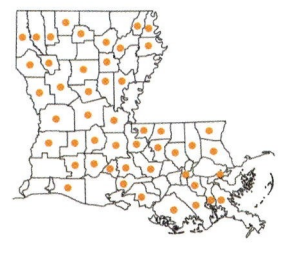

in LA are the Painted Lady and American Lady, and all three act similarly. It feeds at many kinds of flowers (such as Chickasaw plum in the Tunica Hills region) and visits wet spots and puddles, scat, and even the occasional dead snake in the road. Males perch and are wary. Both sexes have a fast, erratic flight but are approachable when feeding at flowers. When disturbed, the Red Admiral flies away quickly, moving in a circular fashion until back at the point from which it was disturbed, as if to find out what had disturbed it. Even though present from late Jan into Nov, this butterfly is rarely abundant.

My Records: I regularly see it throughout the state. Indian Bayou WMA is a good location, where it can be found on the road along the levee and along the numerous four-wheeler trails. I have seen it regularly at Peveto Woods, an odd location that illustrates this butterfly's adaptability. I usually see no more than one or two at a time; however, in Oct of 1993, large numbers were seen on blooming boneset at Whiskey Bay in Sherburne WMA. I've never seen those kinds of numbers since.

Distribution and Abundance: In CAD, it flies from late Feb to early Dec, most common in the spring and then again in the fall. During the Cajun Prairie survey, it was recorded from Mar through Nov, with a high of fourteen in May. Moving east, this butterfly has been recorded at Grand Bay and Sherburne WMA (mid-Nov). In the Felicianas, there appear to be three broods from late Feb into Dec, over which it can be common. In the FL Parishes, it has shown up at Honey Island Swamp WMA (Sept). It has also been sighted at Edgard in SJB. It has regularly appeared during the Pearl River count, occasionally in double digits. In the lower parishes, there are records for Grand Isle, Galliano, Golden Meadow, and Larose for every month, with an earliest of January 10 and a latest of December 27.

Host Plants: Its primary larval food plant is nettles. In southeast TX, it is also reported to use false nettles.

BRUSHFOOTED BUTTERFLIES (Nymphalidae)

Painted Lady (*Vanessa cardui*)

ventral

dorsal

TRUE BRUSHFOOTS (Nymphalinae)

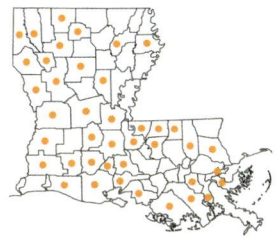

Description and Behavior: Although it is sometimes called the Thistle Butterfly, I have no recollection of ever seeing this butterfly on thistle. It is a denizen of open, disturbed areas and readily visits flowers. These visits yield the best opportunity to observe it; otherwise, it is wary. A swift flier, if disturbed it will often swirl around for a bit and then find another flower close by and start feeding again. I've read of huge migrations by this butterfly in the West, and some sources have suggested it migrates into LA from Mexico and/or TX, but I have not seen or read reports of any such migrations.

My Records: I've seen it in every month but Jan and Feb, in various kinds of habitats. It is not reliable at any particular location. In 2012, none were seen until mid-Sept during a count at Duralde Prairie.

Distribution and Abundance: This butterfly is not particularly common in LA. In CAD, it has been reported in small numbers in May, June, and July with better numbers in Aug through the end of Nov. It was found in small numbers during the survey of the Cajun Prairie region in Jan to Apr, Aug into Sept, and Dec, with a high of thirty-nine in Oct. Turning east, it was recorded in mid-Nov at Sherburne WMA. Records reflect it to be uncommon in the Felicianas from Mar into Dec. During Israel's Tunica Hills survey, sightings were sporadic (Mar, July, Sept to Nov with three broods) in open fields. It turned up once on the Bonnet Carré count, and has been recorded on ten of the Pearl River counts. In southernmost LA, there are records sporadically throughout the year, including Jan and Dec.

Host Plants: It has been reported to use thistles, nettles, althea, citrus, elm, and American elm in LA. In southeast TX, it is reported to also use mallows.

BRUSHFOOTED BUTTERFLIES (Nymphalidae)

American Lady (*Vanessa virginiensis*)

ventral

dorsal

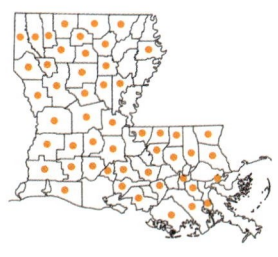

Description and Behavior: Of the two Ladies, this is the more common in LA, present from late Jan to mid-Nov, but more common in the spring into early summer. There are several easy ways to differentiate it from the Painted Lady. The Painted Lady is larger, and its forewings are extended, while this Lady's forewings are more blocked. There are two large spots on the ventral hindwing of this one and five small spots on the Painted Lady. Finally, the pink on this Lady's ventral wings is deeper, contrasted with generally a paler pink on the Painted Lady. Like the Painted Lady, it is a fast flyer and very fond of flowers. It seems to like white flowers, and I've found it on blackberry blooms, ligustrum, and Chinese privet. It also visits clover (purple and white) and dandelions (yellow).

My Records: I see this one more reliably each year than the Painted Lady. I've seen it at Thistlethwaite WMA regularly, but the most dependable place I've found it is in the GRA unit of Kisatchie NF, at blooming blackberry vines along the road that fronts Stuart Lake. The latest I've seen one was at Cypremort Point SP in mid-Nov.

Distribution and Abundance: In north LA, records in CAD are from late Feb to mid-July, common to abundant in late May to mid-June, a single sighting in Sept, then late Oct into Nov. In the central LA region, sightings include VER (June and Aug) and CALD (May). During the Cajun Prairie survey, it was seen from Mar to Sept, peaking in May. Sightings in the Felicianas have been year-round, where it was uncommon (except in Apr to June, when it was common). It has been recorded during the Pearl River count, but in fewer numbers and on fewer occasions than the Painted Lady. In southernmost LA, there are records from Mar through June and Oct.

Host Plants: In LA, records include thistle, nettles, everlasting, althea, and species of Gnaphalium. In southeast TX, it is reported to also use cudweed and pussytoes.

BRUSHFOOTED BUTTERFLIES (Nymphalidae)

Red-spotted Purple (*Limenitis arthemis astyanax*)

ventral

dorsal

"rubidus" hybrid

Description and Behavior: This is a woodlands butterfly, found along the fringes and trails in deciduous forests, flying even in the forest depths. The males can be found basking in sunny spots on dirt roads, trails, or overhanging branches and will fly out to investigate objects that invade their territory. Males are notorious group puddlers, resting with their wings held flat to the ground. I once counted fourteen at a puddle party (with eleven Question Marks) at the boat launch at Bayou Cocodrie NWR. They are also fond of mammal scat and, on rare occasions, will visit flowers, mostly white flowers like boneset. The females like the deeper woods, but can also be found basking on branches extended into the sun. When disturbed, they move away in a flat, flap-and-glide manner, often returning to the same spot in a few moments. If harassed, they fly swiftly up and into the forest.

As a Pipevine Swallowtail mimic, both sexes have black dorsal forewings and blue dorsal hindwings, but no tails. The males are smaller and have bright blue scaling on their hindwings. The females can be quite large during the late summer, with a much duller, blue sheen. In late July, during the 2001 Indian Bayou WMA count, Ross and Marks (2002) caught a hybrid, *L. archippus x*

Red-spotted Purple (*Limenitis arthemis astyanax*) (*continued*)

arthemis astyanax, taxonomically recognized as hybrid form "rubidus" Strecker. Hybrids between these two congeneric species are not uncommon. The Indian Bayou specimen, which more resembled this species, was the third reported from LA. Brou had one in his collection, also taken in July, near Edgard. His specimen was marked more like a Viceroy.

My Records: While I have regularly seen it in central LA's piney woods, it is less numerous there than in the Acadiana region. It is multi-brooded with a spring, midsummer, and fall brood. There have been times at Indian Bayou WMA when it was extremely common with between ten and fifteen seen in a single day, but most times two to five make for a good day. I have seen it only once in the River Parishes (Gonzales) and once in the Florida Parishes (Hutchinson WMA).

Distribution and Abundance: This brushfoot has been recorded around the state, with dates from late Apr into Oct. It is common to abundant in CAD from Apr until Sept, then occasional into Nov, and in the Cajun Prairie area from Apr until Sept. Reported in both of the Felicianas from Mar to Nov, it is most common in Apr, July, and Sept over three broods. It has been seen almost every year during the Pearl River count, with a high of ninety-nine in 2001. Farther south, it has been recorded a few times in LAFO from May to Oct.

Host Plants: Several hosts have been reported: cherry, hornbeam, basswood, black cherry, crabapple, hawthorn, hop hornbeam, oak, pear, white poplar, black willow, and willow.

Viceroy *(Limenitis archippus)*

ventral

dorsal

Viceroy (*Limenitis archippus*) (*continued*)

dorsal "watsoni"

Description and Behavior: At times very common, the Viceroy is multi-brooded (possibly as many as four broods) and typically found near water and willows, its primary host plant. The males perch, flying out to investigate passing objects. Both sexes typically fly in a flat, flap-and-glide fashion about four to five feet off the ground and, if left undisturbed, may return to their previous perch. The Viceroy is attracted to dung.

The subspecies present over most of its national range (*archippus*) is red-orange and mimics the Monarch. The relationship between the Monarch and the Viceroy was first thought to be a Batesian mimicry. However, later studies suggested Viceroys, as willow feeders, were not palatable, thereby creating a Mullerian mimicry relationship. Other experiments have shown that, in the southern US, Viceroys benefited from mimicking Queens, which were more numerous in that region, creating what Cech and Tudor (2005) characterized as a "mimicry triangle" with each a mimic of the other two. In fact, studies by Ritland and Bower (1991) suggest the Viceroy is "nearly as unpalatable as Monarchs and more so than Queens." Israel (1981) reported that during his thesis survey, the flight period of Viceroys essentially matched the Monarch's flight period. The two maintained approximately equal frequency numbers during

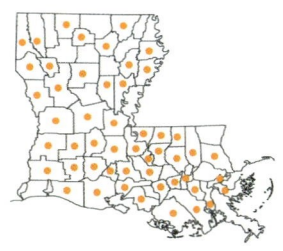

the typical Monarch migration to and from Mexico; however, Viceroys were about twice as common during the period between the migrations.

There are two Viceroy subspecies in LA, *archippus* and the darker *watsoni* (dos Passos). In north LA, I have seen only the former, while the latter subspecies is present in central and south LA. On numerous occasions in the Acadiana region, I have found extremely dark Viceroys. At Thistlethwaite WMA, Indian Bayou WMA, and Avery Island, more than 50 percent of the populations possess *watsoni*'s extremely dark dorsal and ventral forewings. These fly alongside the lighter-colored version. Monarchs are much more common than Queens at Thistlethwaite WMA (I've seen as many as thirty-three Monarchs one day in Apr), so logic would suggest the Viceroys in this region would be mimics of the Monarch. Yet there is a significant percentage of Viceroys that are dorsally dark like Queens. With Queens present in the area, I now wonder if the Viceroys at Thistlethwaite WMA are mimicking both Monarchs and Queens. In fact, the Tvetens (1996) suggested that all three species potentially "receive some protection by adopting the stereotypical orange-and-black pattern that advertises unpalatability in nature." If that is so, whether they are bright orange or burnt orange, the survival chances of Thistlethwaite WMA's Viceroys are increased due to the presence of both milkweed-feeding Danaids, and vice versa.

My Records: There have been times at Indian Bayou WMA, Thistlethwaite WMA, and near the Acadiana Nature Walk where multiple males were perched about five to ten feet apart, with several continually engaged in aerial battle anytime one moved. I have found the dark subspecies as far south as the Maurepas Swamp WMA in early May and as far north as Catahoula NWR in June.

Distribution and Abundance: Recorded throughout LA, Viceroys can regularly be seen from Apr until Oct, even into Nov if the weather remains warm. It was the ninth most common butterfly in the Cajun Prairie survey. During the Tunica Hills survey, it was most common in June to Aug and Oct.

Host Plants: In LA, this butterfly uses willows, bilberry, crabapple, gopher apple, hawthorn, pear, white poplar, plum, and black cherry trees.

BRUSHFOOTED BUTTERFLIES (Nymphalidae)

Common Mestra (Mestra amymone)

ventral

dorsal

Description and Behavior: This small, whitish south TX endemic is rare in the Houston area, but will undergo, on occasion, northward migrations in large numbers, reaching places far beyond its expected range. Although less than half the size of a Zebra Longwing, it flies in a very similar fashion with frequent, shallow wing beats. Typically, it flies close to the ground in open areas along the edges of thick woods or brush. If disturbed, it quickly flies into the woods or brush. If left alone, it will visit low-growing flowers and perch in the sun. The males appear to patrol in a manner similar to male Falcate Orangetips.

My Records: I've seen this dainty butterfly in south TX, as far north as the Corpus Christi area, but not in LA.

Distribution and Abundance: Older records include twenty specimens from CAD during Sept, Oct, and Nov 1957. More recently, one was reported in the Shreveport area in Oct 2007. That same year, another was seen (no photo or specimen) at Cane's Landing in mid-Aug.

Host Plants: The plants used in TX are members of the noseburn family.

Goatweed Leafwing (*Anaea andria*)

ventral male

dorsal female

Description and Behavior: This is a common LA butterfly in the spring and early summer, slightly more abundant to the north. The preferred habitat is open pine forest with areas of secondary growth, where its food plant, croton, grows best. It is less common, but can still be present in a deciduous hardwood-

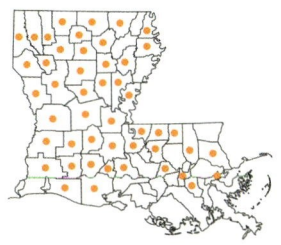

forest habitat. Only slightly dimorphic, dorsally, the female has a broad dark brown border on both wings with additional dark markings interior of that border. The males have a smaller and lighter margin and no other markings. In both sexes, the spring brood has longer tails and a more arched, pointed forewing tip.

This butterfly is most often seen along trails or dirt roads. Typically, as I walk along a sunny trail, it will suddenly fly up in front of me with a flash of bright orange-red. Once flushed, it flies swiftly, usually away from what flushed it, but returns a short time later and lands in virtually the same spot as before. Upon landing, it closes its wings, showing only the ventral surfaces, which blend well with the surrounding dead leaves. At that point, it becomes an extremely wary insect, playing a persistent cat-and-mouse game, not allowing a close approach but flying again, only to again land a few feet farther down the trail.

My Records: More common in the spring, I see this butterfly regularly throughout central LA. It can be found along the dirt roads of the WIN (forty-three seen in one day in mid-Mar 2017), NAT, and RAP units of Kisatchie NF and on the trails at Indian Creek Rec Area. It is also a regular, spring and summer, at Copenhagen Hills along the road that leads into that preserve. East of the Mississippi River, I found it along the tree line in the interior of Mary Ann Brown Preserve. It is common at Hutchinson WMA in late Mar.

Distribution and Abundance: There are records from CAD in every month but Feb, most common in Mar. It is common at Bodcau WMA, Cane's Landing, and the Headquarters Unit of Red River NWR in Mar–May and then from Aug to early Nov. In the Cajun Prairie region, records are from Feb through Aug and then in Oct and Nov, most abundant in Mar. In southwest LA, it has been found at Intracoastal City and Palmetto SP in late Mar and early Apr. There are records from all twelve months in WFE. At Abita Springs, records run from early Mar into early Dec over five annual broods at about fifty-four-day intervals. In contrast to records from farther north in the state, Brou (2013a) reported (primarily from traps baited with fermented fruit) a large increase in sightings in the fall, primarily in Oct. It has been reported four times for the Pearl River count, each time as singles.

Host Plants: In LA, members of the croton family, including woolly croton or goatweed, are reported to host this butterfly.

BRUSHFOOTED BUTTERFLIES (Nymphalidae)

Hackberry Emperor (*Asterocampa celtis*)

ventral male

dorsal male

dorsal female [bleached]

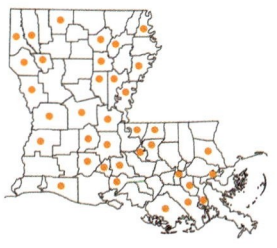

Description and Behavior: This emperor is strongly associated with hackberry trees but can often be found "puddling" at wet spots in the road. I don't see it at flowers, but it will readily visit roadkill, mammal scat, and ultraviolet light traps (see Brou's report [1974] of catching 191 in SJB). It will also land on people, attracted to sweat. I've had them land and then remain on me while I walk several hundred feet. I had one stay even when I got in my car. During the summer, their coloring can become very light, causing them to appear almost ghostlike as they flit through the forest. This "bleaching" may be a result of extended exposure to sunlight while basking/perching or unusual climatic conditions such as an unseasonal heat wave at a sensitive time in development, producing a "washed out" form. These bleached specimens are significantly more prevalent at Indian Bayou WMA than elsewhere. Lambremont (1954) reported that both of the nominal subspecies *A. c. celtis* and *A. c. alicia* had been found in the state.

My Records: I have recorded it from Apr into Oct. This butterfly can be extremely numerous at both Indian Bayou WMA and Thistlethwaite WMA, particularly during the summer after several days of good rain. It seems to become reduced in numbers to the north, becoming less common in the predominantly pine forests of Kisatchie NF.

Distribution and Abundance: A common butterfly in the right habitat, it has been recorded from Mar to Nov. In CAD, it has been seen from mid-Apr until late Oct. It was recorded in the Cajun Prairie area in modest numbers between May and Nov. In the FL Parishes, it was found near Erwinville in June and there are records from both Felicianas, May to late Oct over three broods. It has been a regular during the Pearl River count with a high of 63 in 1993. Farther south, in LAFO and TER, it has been reported from Mar into Oct. Other records from this region were mostly in the Galliano area with a few from Grand Isle.

Host Plant: In LA, as throughout its range, it uses hackberry trees.

BRUSHFOOTED BUTTERFLIES (Nymphalidae)

Tawny Emperor (*Asterocampa clyton*)

ventral male

dorsal male

dorsal female

ventral female

ventral female ("flora" form)

Description and Behavior: The lifestyle and habits of this butterfly are virtually identical to those of the Hackberry Emperor. In many instances, it can be found flying with Hackberries, but typically it is not as common. The females are larger and lighter colored. In mid- to late summer, the females can be quite large, as big as Question Marks. The females 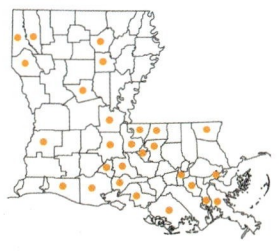 are more reclusive, seen mostly along tree lines, perched on outstretched branches in the sun. If disturbed, they retreat back into the deeper woods.

My Records: My sightings have been between May and Sept, with only two sightings in Oct, both in southwest LA. I see it regularly at Indian Bayou WMA. In the Acadiana region I have noted that the summer and fall flights of this butterfly closely resemble the *flora* subspecies of west FL. This subspecies is orange-red, rather than the darker *clyton* subspecies. Like the Hackberry Emperor, I don't see this emperor as often in the central LA region.

Distribution and Abundance: It is rare but occasional in CAD from Apr through Sept with a single sighting in Nov. In central LA, several have been seen during Copenhagen counts (June and July). Brou had specimens from BEAU in his collection. In the FL Parishes, Strickland (1972) found this butterfly near Erwinville and Grand Bay. In WFE, records are from early June to mid-Oct, over three broods, described as local in association with hackberry trees. It has also been reported as uncommon at Asphodel Plantation from May to Oct. In a coincidence that I can't really explain, the three butterflies that Brou found most commonly in his ultraviolet light traps in SJB were all hackberry feeders: Snouts, Hackberry Emperors, and Tawny Emperors. This emperor has turned up at the Bonnet Carré count and four times during the Pearl River count. Farther south, this emperor has turned up as often as its cousin, primarily in the Galliano area but also at Grand Isle, flying in Apr through Oct.

Host Plant: Hackberry trees.

BRUSHFOOTED BUTTERFLIES (Nymphalidae)

Northern Pearly-eye (*Enodia anthedon*)

ventral

ventral

anthedon antennae

Description and Behavior: Historically, the Northern and Southern Pearly-eyes were considered to be well differentiated but clinal races of a single species, the Pearly-eye, *Lethe portlandia*. By the mid-1970s, the two were reported as separate species by Heitzman and dos Passos (1974), with the Northern Pearly-eye determined to use a variety of woodland grasses as its host plant, and not cane. In my experience, the Northern's coloring is darker and more precise. The wings are smaller and more blockish. Unlike LA's other two pearly-eyes, this butterfly will regularly land on the sides of trees. I had always understood that butterflies don't hear noises; however, some members of the satyr family, including this one, possess auditory sensors within the swollen veins on the forewing base. These hearing organs allow the butterflies to hear approaching predators.

My Records: In early June 2012, at Sicily Island Hills WMA, a photograph by Trahan of one of several satyrs, all tentatively identified as Southern Pearly-eyes because they appeared to have yellow-tipped antennae, was submitted to BAMONA. As part of the verification process, Ricky Patterson (from MS) concluded the photo reflected a Northern, not Southern Pearly-eye. Per Patterson, some Northerns have orangish antennae on the underside, and antenna coloration is not 100 percent accurate as some Northerns can have an orangish antenna with little indication of the darker club, especially on the underside. Distinctive features of Northerns include cleaner wing maculation than LA Southerns. Also, the row of eyespots on the underside of the forewing on Northerns is in a straight line, whereas in Southerns that row of eyespots is curved. Finally, the ventral eyespots on the lower wing of Southern females show an orange coloration, missing on Northerns.

I returned to Sicily Island Hills WMA in late July 2012 to investigate further. There were numerous pearly-eyes present, and specimens were taken. All were active in the shade, flying on several fairly steep slopes where both cane and the possible grass host plant for Northerns were growing. After close inspection of the specimens collected in July, I determined that both Northern and Southern Pearly-eyes are present at Sicily Island Hills WMA. Dorsally, the antennae of the prospective Northern specimens showed the dark clubs. A comparison of several specimens that were clearly Southerns, caught in a separate region of the WMA, reflected the differences Patterson had outlined (see photos). These differences include size, coloring, markings, and, to an extent, habitat.

Distribution and Abundance: Prior to its being located at Sicily Island Hills WMA, the Northern Pearly-eye was listed on BAMONA as part of LA's fauna,

BRUSHFOOTED BUTTERFLIES (Nymphalidae)

Northern Pearly-eye (*Enodia anthedon*) (*continued*)

with a record from WFE without particulars—sources, dates, locations, and so forth. Brou (2013c) reported that he found this butterfly, a male, in EVA in mid-July 1979. Patton reported on the LA listserv that he had found another at Chico SP in that parish in Aug 2015.

Host Plants: In AL, it is thought to use woodland grasses such as river oats, white cutgrass, and eastern bottlebrush grass. Charles M. Allen, author of the book *Grasses of Louisiana,* identified the grass at Sicily Island Hills WMA as either slender woodoats or longleaf woodoats, two closely related grasses, typically present along the edges of and within forests. Neither has been specifically identified as a host plant, but several authors have generically listed "forest grasses," including sister species such as Indian woodoats, which grow in shaded, wooded habitat.

Southern Pearly-eye (*Enodia portlandia*)

ventral female

dorsal female

Description and Behavior: Fond of moist, shaded forest roads and power-line cuts, this pearly-eye can be found at mud and dung. It flies low to the ground in a fast, weaving fashion. It is always found near cane, but not necessarily near water. It will stop and perch, often on leaves of cane, or on leaves of grass near the cane. A quick field method to identify pearly-eyes is to check the end of the

BRUSHFOOTED BUTTERFLIES (Nymphalidae)

Southern Pearly-eye (*Enodia portlandia*) *(continued)*

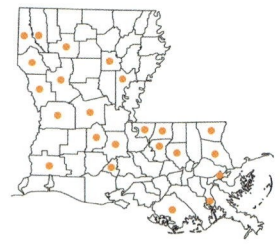

antennae. If the last few millimeters of the tips are completely orangish-yellow, it is a Southern, but if the tips are dark, it is not. A large satyr, the female of the species is usually larger than the male. It is multi-brooded, flying from early spring to late fall, common at times, virtually absent at others. Brou (2013c) reported six annual broods at forty-five-day intervals in the Abita Springs area, while Israel (1981) reported three in the Tunica Hills region. At Abita Springs, Brou was finding it at traps baited with fermenting fruit.

My Records: It can be quite common at Avery Island and regular at Thistlethwaite WMA. The population at Avery Island is more yellow or blonde dorsally than specimens found in more northern sections of LA. In central LA, I have found it in the NAT Kisatchie unit in one of the hardwood bottoms west of the Caroline Dormon Trail (Oct), at Sicily Island Hills WMA (Mar, June, July), and at Copenhagen Hills (June). In the FL Parishes, I have found it in Mary Ann Brown Preserve between Apr and Sept, during multiple months in consecutive years at Tickfaw SP, and in late Apr and late Sept at Abita Creek Flatwoods Preserve along the boardwalk. In all of these locations, it was within sight of cane.

Distribution and Abundance: Skinner reported "*Lethe portlandia portlandia*" from SLA with a sighting/capture date in July 1897. Since then, it has been recorded flying from Feb through Dec and described as abundant at times in the right habitat. In north LA, locations include Eddie Jones Park (Oct and Nov), Bodcau WMA (Apr through June and Aug), and Madden Mill Road (June). In central LA, it has been found in Kisatchie NF in RAP (Mar and Aug). In the Cajun Prairie region, records reflect it flying in Mar, May (only one seen), July (again, only one), then Aug into Nov. It has also been found in CALC. In the FL Parishes, it was recorded in lowlands of EBR, including Greenwood Park, into Nov and Dec. It can be common at Asphodel Plantation and in WFE, Mar through Nov, most common in Apr and Oct. These records also include several Pearl River counts (thirty-six in 1999). Farther south, it has been reported during the 1998 Barataria count (June) and in TER.

Host Plants: Giant cane has been identified as the food plant in the Felicianas. In southeast TX, the larval food plant is reported to include switch cane.

Creole Pearly-eye (*Enodia creola*)

ventral female

Description and Behavior: This pearly-eye is closely associated with, but much rarer than, its host plant, switch cane. It is almost always found flying with Southern Pearly-eyes, but in fewer numbers and locations. It is extremely difficult to tell the difference between the two species in flight as both move with a weaving flight, swifter and less bouncy than other southern satyrs. The males are easy to identify as their forewings are more elongated and narrower than Southerns with the forewing tips pointed outward slightly. The female looks more like a Southern, but will have a fifth eyespot on the ventral forewing. Further, the amount of white scaling around the eyespots on the ventral forewing differs between Southerns and Creoles. On Southerns, the white tends to completely surround all of the eyespots. In contrast, on the Creoles, the white tends to resemble rings around the individual eyespots. At Sicily Island Hills

Creole Pearly-eye (*Enodia creola*) (*continued*)

ventral male

WMA, where all three pearly-eyes are present, identification is more compli-cated. On all three, there is a dark line that runs approximately perpendicular to the leading edge, about halfway between the body and the forewing tip. On Southerns, that line is virtually straight. On Northerns, it is more curved, while on Creoles it is wavy, described by some as resembling knuckles. I believe it to be multi-brooded in LA, with possibly four broods—Mar–Apr, May–June, July–Aug, and Sept–Oct.

My Records: The two most reliable places to find this pearly-eye are Tickfaw SP (Apr to May, Aug to Sept) and Clark Creek Nature Area in MS, just across the state line from WFE (Mar to Apr, June to Oct). I have found that males at Tickfaw SP will land on the trail, whereas the females seem to stay in the woods. Both are difficult to approach. At Clark Creek Nature Area, the Creoles like to hang out (literally) in the vegetation along the vertical sides of creek beds or deep in the cane on which they feed.

Distribution and Abundance: This species was listed in 2015 as a Tier II, S3-ranked SGCN in LA. First reported in LA by Skinner in 1897, that specimen was later determined to be a Southern, not Creole, Pearly-eye. In Acadiana, one was reported at Washington in Oct 2008 (without photos) and at Chicot SP in Aug 2015. In the central LA region, it has been 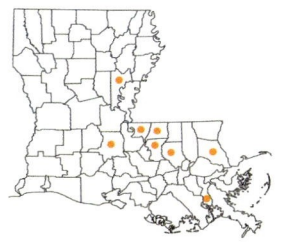 seen at Sicily Island Hills WMA in June. Most state sightings are from southeast LA, such as WFE (Apr and Oct) and EBR (June, July, and Sept), near switch cane. Other sightings include Tunica Hills WMA, in Mar. Israel considered it rare at Asphodel Plantation from Mar into June and then from Sept to Oct. It has been reported during a couple of NABA counts, Pearl River (June 2005 and 2007) and Barataria (June 2004).

Host Plant: Switch cane.

BRUSHFOOTED BUTTERFLIES (Nymphalidae)

Appalachian Brown (*Satyrodes appalachia*)

ventral

dorsal

Description and Behavior: Historically, this butter-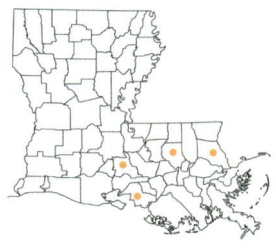
fly was considered to be the southern form of the
Eyed Brown (*S. eurydice*) until recognized in the
1970s as a separate species. Despite the name, its
recognized range extends well beyond the Appalachian Mountains southward into FL and west into
AL and north MS. There are also colonies in AR,
but none particularly close to LA. It was listed in 2015 as a Tier III, SU-ranked
SGCN in LA.

My Records: In Aug 2007, while scouting for the Indian Bayou WMA count,
I found an area of moist hardwood bottoms with some standing water and
stands of a tall sedge that is a host plant for Dukes' Skippers. In addition to
several Dukes', I observed a large, satyr-like butterfly flitting among that sedge
that turned out to be an Appalachian Brown. I found another in that same area
about an hour later, the first record of this species in LA west of the Mississippi
River. Not knowing this species as a migrant, I initially assumed there was a
small colony at Indian Bayou WMA; however, I did not see another one at that
location until early June 2015, when I found a tattered female. Multiple specimens were seen at this site in July 2016. In June 2016, while scouting a pipeline
in the North Bend (East) Unit at Bayou Teche NWR (where I had found Dukes'
Skippers the year before), I found seven, males and females, most very fresh,
back in the heavy forest alongside the pipeline. In mid-Sept of that same year,
I saw thirteen fresh individuals flying in that same pipeline.

Distribution and Abundance: V. A. Brou reported (by e-mail) finding it in May
1995 at Abita Springs. Specifically, he found three specimens in his bait traps
on separate days about thirty minutes before dark. Gary Ross (also by e-mail)
reported it in Oct 1981 at Denham Springs. Both records are from the FL Parishes, east of the Mississippi River. Via a post on the LA listserv, Patton reported
finding another back at Indian Bayou WMA in Aug 2015. The combined LA
sightings suggest three broods: May–June, July–Aug, and Sept.

Host Plants: In AL, it is reported to use sedges. The habitat at Bayou Teche
NWR is moist hardwood bottoms with a combination of savannah panicgrass
and Carex sedge growing, with the former predominating. I did not see any
ovipositing, but the specimens I saw were more associated with the panicgrass
than the sedge, flying over and landing on it repeatedly.

BRUSHFOOTED BUTTERFLIES (Nymphalidae)

Gemmed Satyr (*Cyllopsis gemma*)

ventral

dorsal

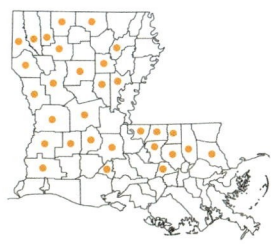

Description and Behavior: This satyr is more common than the Georgia Satyr but not as common as the Carolina Satyr. It is a creature of the forest understory, flying in the shade of upland pine habitats and hardwood bottoms. It is typically larger and a lighter gray than the Carolina Satyr. Its flight is low, but stronger and more direct than the Carolina. It is multi-brooded, flying from early Mar into Nov, including all months between.

My Records: Mary Ann Brown Preserve in WFE is a good place to see Gemmed Satyrs, on the Main Trail Loop at the back of the preserve. The grass on which I believe this butterfly feeds grows with cane at that location. I can also usually find it at Thistlethwaite WMA in a wooded area near a slow-moving slough. Although the area is wooded, there is a good amount of grass that I believe is the larval food plant growing under the trees. In both locations, Southern Pearly-eyes and Little Wood Satyrs also are present. By walking through these two areas, sweeping the grass with a net, I have found that the satyrs will fly up from where they were perched.

Indian Creek Rec Area is one of two other locations where I have found all three of the small satyrs (Gemmed, Carolina, and Georgia) to be present, although I've not found them flying there at the same time or same place. The other location was along the Whiskey Chitto River in ALL, near Oberlin. That river is popular for canoeing. In mid-June 1997, my family made the trip, stopping for a picnic lunch along the way in a location that I could not now find if my life depended on it. While they ate and rested, I walked into the forest and quickly found all three satyrs flying together.

Distribution and Abundance: This satyr has been recorded from across the state with an overall distribution similar to the Little Wood Satyr. It is occasional in CAD from late Feb into mid-Nov. It has also been reported at D'Arbonne NWR. Small numbers were reported in Mar, May, and June during the Cajun Prairie region survey. It has been seen in both of the Felicianas, Mar to Nov over three broods, found flying with other satyrs. It has been reported once on the Global Wildlife count (July) and five times during the Pearl River count.

Host Plants: In southeast TX, the larval food plant is reported to be Bermuda grass with other grasses likely. In AL, it uses shade-tolerant grasses like slender spikegrass.

BRUSHFOOTED BUTTERFLIES (Nymphalidae)

Carolina Satyr (*Hermeuptychia sosybius*)

ventral

dorsal

Description and Behavior: The most common satyr in the state, it, along with the Pearl Crescent and Common Buckeye, are LA's most common butterflies. It is found during all months. There have been some trips during the humid heat of Aug when this satyr was the most common butterfly seen, flying in the relatively cooler shade. The Carolina Satyr is smaller than the other two small satyrs, darker gray, and flies slower. It flies within the forest's understory, low to the ground. It will fly up into the trees if roughly disturbed, shortly to drop back down to ground level. When two of these satyrs meet, they start a tight, whirling, twirling dance just above the grass, which typically ends with the two headed off in opposite directions. It is multivoltine.

A short story: There was a Chinese privet tree, between fifteen and twenty feet tall, in the right-of-way along the main road into Thistlethwaite WMA. It would bloom during the last two weeks of Apr and attract hairstreaks like bees to honey, so I called it the hairstreak tree. It was a guaranteed location to find Banded, Southern, and Red-banded hairstreaks. It also attracted Silver-spotted Skippers, Buckeyes, American Ladies, Red Admirals, and Pipevine Swallowtails. On more than one occasion I found Seminole Crescents, Snouts, Hackberry Emperors, and even Little Wood Satyrs on the flowers. Even more odd (and the point of this story), each year I would also find Carolina Satyrs taking nectar on this tree just before dusk, the only time I recall seeing this butterfly feeding at flowers. Alas, for reasons that escape me as it did not impede the road's right-of-way in any fashion, someone decided to cut down the tree in 2011.

My Records: This is one of the three species that I consider to be the most common butterflies in LA. I find it everywhere there are woods, be they deciduous or coniferous. I've found it flying with Georgia Satyrs at Indian Creek Rec Area, Cooter's Bog, Lake Ramsey Savannah WMA, and Abita Creek Flatwoods Preserve.

Distribution and Abundance: It has an extended range all around the state. In CAD, it is abundant, flying from late Mar into mid-Nov. In the central LA region, it is very common, with records from May to Sept. During the Cajun Prairie survey, it was the tenth most reported butterfly, with records from Feb (one seen) into Nov (also one seen), most common in Apr. In the Felicianas, it is reported as always common and often abundant from late Jan into early Dec, with three broods, even flying out in open, grassy areas on cloudy days. In late Sept 2013, 323 were counted at Honey Island Swamp WMA. Although the species is not known to visit flowers, Brou (1974) reported catching 17 at Edgard

Carolina Satyr (*Hermeuptychia sosybius*) (*continued*)

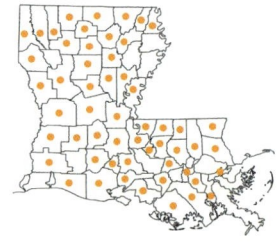

in ultraviolet light traps. In southernmost LA, there are records for all year.

Host Plants: Generically identified as grasses, in southeast TX, the larval food plant is reported to be Bermuda, carpet, and centipede grass. In south FL, St. Augustine, signalgrass, and spikegrass are also listed.

Note: In 2014, Cong and Grishin suggested the existence of a sympatric and synchronic species with *sosybius*, which they identified as *H. intricata*, given the common name of the Intricate Satyr. The authors "were not able to find reliable wing pattern characteristics to tell a difference between the two species." However, the two differed in their genitalia and DNA sufficiently to justify, per the opinions of the authors, separate species status. The initial specimens were described from Brazos Bend SP in east TX, and the distribution was described as the southeast coastal plains. The article listed two specimens of this new species from LA, in JEF (Harahan) and JAC (Jonesboro). As it would be impossible to do so based on data available to me at this time, I make no separation between these two sympatric species.

Georgia Satyr (*Neonympha areolatus*)

ventral

dorsal

BRUSHFOOTED BUTTERFLIES (Nymphalidae)

Georgia Satyr (*Neonympha areolatus*) (*continued*)

Description and Behavior: Of the three small satyrs (Gemmed, Carolina, and Georgia), this is typically the largest, with some females as large as Little Wood Satyrs. This satyr is also the lightest colored of the three (the Carolina Satyr is the darkest). This satyr is often found out in open grassy areas in full sunshine while the other two rarely leave the shade of the forest. The late Ron Gatrelle (1999) proposed the existence of two sister species, *N. areolatus* and *N. helicta*. *Helicta* was defined as an upland butterfly with rounder hindwing spots encircled in orange. *Areolatus* was its lowland counterpart with elongated eyespots.

My Records: During May 2007, while adjacent to Indian Creek Rec Area and within Alexander SF, I found an extended area of mostly open pines with some hardwood trees. There were large patches of high grass and several seeps with just enough water to get my feet wet. I ended up finding numerous specimens that were consistent with *areolatus* with many large females. I was so engrossed in counting these satyrs that I got hopelessly lost for over an hour, as a result of which I now wear a compass on my wrist, compliments of my son and oldest daughter (given to me to avoid the embarrassment of having to explain how their father got lost "chasing butterflies"). When I returned in Aug, the "seeps" had dried, but the satyrs were still there, in fewer numbers. In 2010, that particular area had been burned as part of a prescribed burn, and I have not seen the Georgia Satyr there since.

Two of the three locations where I've seen this satyr are west of the Mississippi River. In addition to Indian Creek Rec Area, I saw more than sixty once at Cooter's Bog in mid-May and more than forty in late Aug, with smaller numbers seen in mid-June and early July. East of the Mississippi River, I found ten at Lake Ramsey Savannah WMA in Aug and eleven at Abita Creek Flatwoods Preserve in late Sept. It was still flying in good numbers at both locations in mid-Oct of that year, although those flying in Oct were noticeably smaller than the June and Aug broods. The males, in particular, were about the same size as Carolina Satyrs. I also found this satyr near Enon in WAS in Aug. Those that I have found east of the Mississippi River are also *areolatus*.

Distribution and Abundance: Historically reported only from the longleaf pine flats north of Lake Pontchartrain in the FL Parishes, it was considered to be rare west of the Mississippi River. More recent records have shown it to be equally present in the piney woods of central LA, including the NAT and GRA units

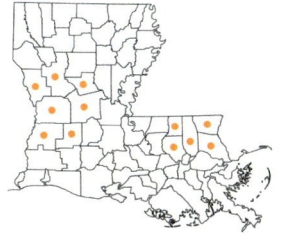

of Kisatchie NF. Fourteen were reported during the Fort Polk count in Sept 2012. It was reported as common in the Cajun Prairie area, flying from Apr to Oct, with peak flights in May–June (forty-one in May, fifty-four in June) and Sept–Oct. Back in the FL Parishes, newer records include SHE during mid-May in open, disturbed pine forests with a heavy cover of grass. It has been reported twice on the Global Wildlife count (July 1992 and 1994). It was listed in 2015 as a Tier II, S3-ranked SGCN in LA.

Host Plants: In southeast TX, the larval food plant is reported to probably be sedges. In AL, it uses the sedges that grow commonly in pine savanna understory.

BRUSHFOOTED BUTTERFLIES (Nymphalidae)

Little Wood Satyr (*Megisto cymela*)

ventral

dorsal

Description and Behavior: This is the largest satyr in LA. Its low, bouncing flight and habitat preferences are similar to those of the Carolina and Gemmed Satyrs. While seen primarily within the forest, it will venture out into open, grassy areas along the forest's edge. When disturbed, it accelerates, bobbing and weaving through the understory and quite hard to follow.

Some sources suggest *M. viola* may be a separate species. Others list it as a subspecies. It is characterized by its larger size and larger, brighter eyespots. Its range is described as along the lower East Coast into central FL as well as the FL Panhandle. There has been a suggestion it may occasionally extend into LA, but without substantive support. While the May brood I typically see is both larger and "more spectacularly spotted" than the earlier brood, with so much uncertainty about what is and is not *viola*, I draw no conclusions for LA.

My Records: My records are primarily from Mar to May, and indicate that the species is common to abundant when in flight. I find it regularly at Thistlethwaite WMA and Mary Ann Brown Preserve, where, at the right time, it can be the most abundant butterfly seen. I have also seen it in good numbers at the RAP and GRA units of Kisatchie. It flies at the same time and place as King's Hairstreaks at Blue Hole Rec Area. It was common at Hutchinson WMA in Mar and at Northlake Nature Center in mid-Apr. I have June and July records, but never in the numbers that fly between Mar and May. I found it twice at Copenhagen Hills in mid- to late June. I believe there are multiple broods in LA: a flight in late Mar into Apr, a second in May into early June, and a third (partial?) in July and later. I do not know if the second flight is the progeny of the first flight or from the previous year; however, the second flight is typically larger. It has been recorded on both the Indian Bayou WMA and Shreveport counts in July, with three seen at each. It was also seen on the 2012 Fort Polk count in Sept.

Distribution and Abundance: Older records were only from early Apr through late June. It has been found throughout most of the state, although I have found no records from the marshy southeast portion of LA. It is common to abundant in CAD from late Mar into late June, then with individual sightings in Sept and Oct. I found one record from the Kisatchie unit in RAP in Aug. It was recorded in good numbers during Mar through June in the Cajun Prairie region. In the Felicianas, Israel (1981) reported it to be locally common over two broods from Mar into June. Its preferred habitat was described as "lighter areas of the forest

BRUSHFOOTED BUTTERFLIES (Nymphalidae)

Little Wood Satyr (*Megisto cymela*) (*continued*)

where grasses grow on the forest floor." In southeast LA, it has been reported on the Pearl River count in June on three occasions.

Host Plants: In southeast TX, the larval food plant is reported to be grasses like orchard grass and centipede. In south FL, St. Augustine grass has been listed.

Common Wood-Nymph (*Cercyonis pegala*)

ventral

dorsal

BRUSHFOOTED BUTTERFLIES (Nymphalidae)

Common Wood-Nymph (*Cercyonis pegala*) (*continued*)

Description and Behavior: Lambremont (1954) said it best: "The insect usually may be found sitting on the sides of pine trees, and unless disturbed it is often mistaken for a loose piece of bark. The undersides of the wings are colored in such a way that they blend with the surroundings so well that one must 'beat' the trees with a stick to flush them from hiding. When disturbed the butterfly quickly flies a short distance to another tree, and the same procedure must be repeated." I would add only that, if you stay after them, they will ultimately fly into the grass to hide, thereby actually becoming easier to approach. Females are larger and lighter colored than males. Neither sex readily visits flowers, and Brou's studies suggest they are attracted to fermenting bait.

Brou (1993) recorded the subspecies he studied in STA to be *C. p. abbotti*; however, more recently, Calhoun (2017) reported that subspecies, originally described from specimens found in northern FL and southern GA, more properly reflects a cline between the "northern and southern extremes" of *C. p. pegala*. Calhoun further suggested a "fairly narrow blend zone between *C. p. texana* (present in Texas) and *C. p. pegala* [which] occurs along the Gulf Coast, mostly in LA." Calhoun reported that males in FL have one eyespot on the forewing. I have regularly found specimens with one eyespot in the yellow field on the upper wings along the Gulf Coast in AL and MS, as well as in STA, but at a much-reduced frequency west of the Mississippi River.

My Records: I have primarily found this butterfly in central LA. I found it in great numbers at Indian Creek Rec Area in 2009 and sporadically since. In the NAT unit of Kisatchie NF, I have seen it over several years flying along the Caroline Dormon Trail at Forest Service Road 360. I found it at Cooter's Bog in mid-June, late Aug, and early Oct. The farthest south I have found it west of the Mississippi River was at CC Road Savanna Preserve in southern ALL. The latest I have seen this species (all females) was in the NAT unit in mid-Oct, in open pine forest. In southeast LA, I've found it at Abita Creek Flatwoods Preserve and Lake Ramsey Savannah WMA, near the LA-MS state line. There, as in central LA, the dominant habitat is open pine woods with tall grass.

Distribution and Abundance: This butterfly has been recorded in southeast LA, central LA, and the FL Parishes. It was recorded in the Cajun Prairie survey from June until Sept, with most sightings in June and July. Other records include near DeQuincy (Sept) in CALC and Kisatchie NF in WIN (July and Oct).

In the FL Parishes, Brou (1993) found it from June to Oct in the Abita Springs area. He believed there was an initial brood beginning in mid-June, peaking in late June and into early July, with a second partial brood about seventy days later. In late June, there were several days during which he found between thirty and forty specimens in his bait traps. Other 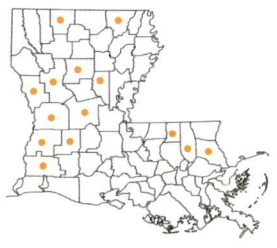 records from that region include SHE in late Mar. NABA count records include west STA (thirty-five in June), Global Wildlife (July and Aug), Fort Polk (Sept), and Pearl River (twelve times with a high of sixteen in 1973).

Host Plants: In the South generally, it has been reported to use various grasses including redtop, purpletop, wild oats, and several members of the bluestem and switchgrasses families (Gramineae).

APPENDIX A

Other Possible Species and Questionable Louisiana Records

OTHER POSSIBLE SPECIES

Lambremont (1954) identified thirty-nine species he "expected" would ultimately be found within the state. (That number was actually thirty-eight because he listed as separate species what are now recognized as the Northern and Southern subspecies of the Oak Hairstreak.) Of that number, thirty-three were indeed found here. (I include *E. bayensis*, which he called *Atryone alabamae*.) Lambremont included *Hesperia attalus* on his "expected" list. That skipper has been reported from the state, but I can find no data to substantiate the record. For the reasons listed below, I have included it on my list of doubtful records. Those on his list but not yet found here were *Strymon titus*, *Erynnis icelus*, *Polites peckius*, and *Amblyscirtes belli*. I have included *titus* and *belli* in my list.

Olympia Marble (*Euchloe olympia*)
I've seen this butterfly only once, in southwest AR. I include it here based on records both north and west of LA's northwest corner. In AR, it has been regu-

larly seen at Lake Catherine SP near Hot Springs. Roever has verbally indicated that he believes he saw it in the southern parts of Polk and Montgomery counties. In TX, it has been found near Dallas at the Lyndon B. Johnson National Grassland. To the south and east of Dallas, it has been reported at Gus Engeling WMA. Therefore, it might show up in BOS and/or CAD. Its preferred habitat is varied and includes shale and limestone barrens, prairies, foothills, open woods, and meadows. In flight, it is larger than the Falcate Orangetip, with which it flies in late Mar; however, like the Falcate, it flies low to the ground in open woods. There is one spring flight, late Mar into early Apr. The males patrol, and both sexes stop at flowers. The host plant is small rockcresses.

Soapberry Hairstreak (*Phaeostrymon alcestis*)

This is a large hairstreak. I would describe it as slate gray, similar in coloring to Edwards' and King's hairstreaks. The distinctive field mark is a white bar with black trim located in the center of the forewing and hindwing cells, horizontal to the body. Its primary range is KA through OK into central TX, but it could ultimately show up along the LA-TX border. For example, it has been reported at Red Slough WMA in extreme southeast OK. In the Dallas area, it has been reported at a couple of locations, flying from mid-May to mid-June. To the south, it has been reported from the upper Gulf Coast region of TX at locations such as Baytown and Galveston. I have found it at Brazos Bend SP south of Houston. In that area, it flies late Apr into early May, and is closely associated

ventral

with its host plant, soapberry trees. The males perch on the leaves of that tree, and both males and females feed on the small white flowers the tree produces during the spring. Soapberry trees have been documented in CAD, and the search continues.

Edwards' Hairstreak (*Satyrium edwardsii*)

This is another large hairstreak from the same subfamily as the Banded, Striped, and King's hairstreaks. Although I've seen it only in the Midwest, where it is more common, it has been reported from Village Creek SP in the Big Thicket region of east TX, within about thirty miles of the LA border. It is unknown if this possible colony survived the significant damage wrought by Hurricane Rita. In the Dallas area, it has been found at Lyndon B. Johnson National Grassland.

In Mar of 2000, James Bess with OTIS Enterprises issued "A Final Report on Insect Surveys at Three Arkansas Natural Area Complexes" (the Bess Report), prepared for the Nature Conservancy of Little Rock, AR. That report covered several units within the Alcoa Bauxite Natural Areas in Saline County, the Columbus Prairie Preserve in Hempstead County, and Miller County Sandhills Preserve in Miller County. All three locations are in southwest AR with the Miller Preserve in the extreme southwest corner, just a couple of miles from the northeast corner of CAD and the LA state line. Within that report, this hairstreak was said to be common in May at the Miller County Sandhills Preserve. These locations in TX and AR have good growths of scrub oak, its host plant, which is also common all along the LA-TX border.

There is one feature associated with this hairstreak that is not shared by the other hairstreaks in LA. It is reported to be found in open scrub-oak habitat that

ventral

also contains ant colonies, which tend to the butterfly's larvae. These ants are primarily part of the Formicidae family. Per Cech and Tudor (2005), there are other circumstances that must occur in any one location to support this hairstreak. For example, this butterfly has a short larval stage, and the ants that tend it require the presence of other "honeydew"-producing insects, like treehoppers and scale insects.

Coral Hairstreak (*Satyrium titus*)

This hairstreak has been reported to the west, north, and east of LA. For example, it was reported from Marshall, Calhoun, and Chickasaw counties in north MS in May and June. It was also reported in the Bess Report to be uncommon at Rick Evans/Grandview Prairie WMA, Hempstead County (although we've never seen it there), Miller County Sandhills Preserve in Miller County, and Dry Lost Creek Glade in Saline County in southwest AR. It has also been reported as rare in the Dallas, TX, area at Lyndon B. Johnson National Grassland. It is a large, tailless hairstreak. Although virtually every reference book makes mention of

ventral

its fondness for butterfly weed, I have never seen it at that flower, although I do agree that it is readily attracted to other flowers. It is much more of a generalist than most hairstreaks, using numerous plants as its host plant (wild cherry, chokeberry, and several kinds of plum) while frequenting all kinds of open areas throughout its range.

Hessel's Hairstreak (*Callophrys hesseli*)

This hairstreak is rare throughout its range in the South, limited to swampy areas where its host plant, Atlantic white cedar, grows. The southern subspecies, *C. h. angulata*, is reported from SC, GA, and along the Gulf Coast (where I have seen several in extreme western FL and southeastern AL). In FL and AL, there are two broods, Apr to May and mid-July into early Aug. Typical habitat is Atlantic white cedar stands in swamps, bogs, and along backwater streams. It

ventral

is especially fond of stream corridors where cedar swamps are frequent. It will visit flowers (reported as more prevalent in the mornings) and mud. It has also been reported to use black titi, a shrub that grows around the base of cedar trees and produces white flowers in the spring. Along the Gulf Coast, it has been reported from Baldwin County, AL, and from

Jackson County and Forrest County, MS. There are known stands of white cedar along the extreme southern LA-MS border in the Pearl River WMA. These areas are hard to access but should be searched with this hairstreak in mind.

Silvery Blue (*Glaucopsyche lygdamus*)

I have included this butterfly based on records both north and west of LA's northwest corner. In TX, it has been reported as rare at Atlanta SP southwest of Texarkana. In AR, it is regularly seen at Lake Catherine SP near Hot Springs. Roever has verbally reported it present at other locations in southern Montgomery and Polk counties in southwest AR. There is only one brood, in early spring (late Mar into early Apr). Its preferred habitat is open areas in damp woods. Males patrol, moving in a low, steady flight. Both sexes stop to take nectar at low-growing flowers along the forest floor. They definitely like white flowers. The blue on the forewing of the Silvery Blue is distinctive, darker, and less silvery than the Azures, which fly at the same time. It flies considerably closer to the ground than the Azures, although Azures will come down low to take minerals at puddles. When on the wing, the Silvery Blue seems to be larger than when examined in the net. Look for its food plant, Carolina vetch, as this butterfly doesn't typically stray too far from it.

Baltimore Checkerspot (*Euphydryas phaeton*)

Primarily found in the East, the Baltimore Checkerspot ranges from ME to northern GA, west to MN, and south to east TX. The *ozarkae* subspecies has been reported in north MS from four locations (Grenada, Lee, Tishomingo, and Webster counties), historically found in late May and early June. More recently, Ricky Patterson advised (by e-mail) that this checkerspot had also been found in the Calhoun County WMA and the Chickasaw County WMA. It is sporadic,

appearing in a location for a couple of years only to disappear. I found it twice, both times in mid-May, once in Webster County and then in Chickasaw County.

It is even more regular in AR, in scattered colonies throughout the Ozarks. I see no reason why it could not be present in northeast LA, the least reported area in the state. These butterflies fly and act in a way very similar to the Silvery Checkerspot (*Chlosyne nycteis*), a close relative. Both are colonial; when you find one, chances are good you will find more. Like the Silvery Checkerspot, unless disturbed, the flight is slow and leisurely. The males fly less than a foot over the ground cover in a flap-and-glide fashion, flying back and forth in what is clearly a patrolling action. The females perch on low-growing leaves and are easily approached.

Tropical Buckeye (*Junonia evarete nigrosuffusa*)

My experience with this butterfly has been in south FL and the Caribbean, but it is reported to fly along the Gulf Coast in south TX. It is not reported as part of the Houston area fauna; however, it has been reported as far north and east as Galveston and Baytown in the upper Gulf Coast region of TX, so it should not be completely unexpected if it were discovered in CAM or CALC in late fall. In late Sept 2013, at Peveto Woods in CAM, I saw an extremely dark Buckeye. It had no white/off-white band, and the forewing tips were more pointed, suggesting it was a male. It was very wary. I could not get close enough to take a photo. The third time I approached, it flew off, not to return. I saw it well enough to state that, unless it was an aberrant Common Buckeye, it was the dark or "nigra" subspecies of this butterfly.

Gold-banded Skipper (*Autochton cellus*)

This spread-wing skipper is rare in the eastern portion of its range. It has two broods in FL, Apr–May, and again in late July into Aug. It is rare in MS with records in May and June to Aug from the central and northeastern sections of that state, and locally rare to uncommon across AR, but absent in southwest AR. In TX, it is reported from the Dallas area as very uncommon. Because it has been found in several locations in north-central FL that are in the same latitude as LA, with habitat similar to that found in portions of LA along the Mississippi River corridor, it may still turn up one day in northeast or southeast LA.

The host plant in the East has historically been described as American hog peanut, but at locations in FL and WV it has been reported to use thicket or wild bean, with which it is closely associated. It flies along trails in shaded wooded areas, beneath large, well-spaced trees where heavy understory is lacking. Both sexes perch on larger leaves in spots of sun to bask. They will also occasionally land on the underside of such leaves when disturbed. They are reported to be attracted to forest-floor flowers like the yellow composite flowers of bear's-foot.

ventral

dorsal

Berry's Skipper (*Atrytone berri*)

Berry's Skipper was first described in 1941, named for the FL collector, Dean Berry, who discovered it. Cech and Tudor (2005) have referred to it as "one of the rarest and most elusive of all East Coast skippers, with few known and reliable colony sites." Its range extends into the MS Gulf Coast (Jackson County). Sources suggest two flights, Apr–May and Sept–Oct, most common in Sept. Patterson's records were for the last two weeks of Sept into early Oct, although I once found it in MS in early Sept. Its host plant is generically identified as sedges, and typical habitat has been described as weedy edges around marshy areas such as ponds, swamps, and canals. I have found it in power-line cuts with thick stands of sedges where both sexes fly down in the sedges. Males both patrol and perch. Both sexes are attracted to blue flowers like liatris, thistle, and pickerelweed. Much of the MS Gulf Coast as well as into southeast LA is inaccessible marsh. I would not be surprised if this skipper turned up in the marshes south and east of New Orleans.

ventral female

ventral

dorsal female

Palmetto Skipper (*Euphyes arpa*)

This skipper has been reported along the Gulf Coast of MS and AL. In AL, it has been found in eastern Baldwin County (I have seen it at Fort Morgan). In MS, there is an old record from the Gulfport area (Harrison County) in Sept, and, more recently, it was found near Moss Point and around Bay St. Louis. The best time to search for it is mid- to late Sept along gravel roads and power-line cuts at flowers (preferably purple) as it is an avid flower feeder. It is a large orange grass skipper that inhabits open pine savannas with a palmetto understory. There are two, possibly three, broods spread out from Mar into May and then July to Oct. Males are reported to perch in the early morning and again in late afternoon, with both sexes visiting flowers during the middle of the day. The flight is described as fast and difficult to follow. I can't explain why this skipper has not yet been found in LA. Its host plant is saw palmetto, a plant that is not just common, but actually abundant in many locations in south LA; however, this skipper does not appear to favor habitat with palmettos and a thick, deciduous canopy. Given its presence along the AL and MS Gulf Coast, it could also be in extreme southeastern LA with its open pine-flatwoods habitat.

dorsal male

ventral

dorsal female

ventral

Carolina Roadside-Skipper (*Amblyscirtes carolina*)

This small skipper's range is from southeast VA, south along the coastal plain and piedmont regions of NC and SC, and then west across the northern portions of GA, AL, and MS. As with the Reversed Roadside-Skipper (below), its range is significantly more restricted than that of its food plant, switch cane. It has been reported in the interior of MS, from Apr until Sept (excluding May), and has been shown on some range maps as present as far south as adjacent to the northeast LA region. Its preferred habitat has been described as grassy pine flatwoods, savannas, and the edges of bay swamps and pocosins. Three broods have been reported in NC, from Apr through Sept, appearing earlier in the spring than the other cane-feeding skippers. This skipper is similar to the Reversed Roadside-Skipper, but they can be differentiated ventrally. This skipper has a pattern of pale spots against a yellow background on the ventral hindwing, while the color of the ventral hindwings on a Reversed is darker and less yellow. Also, the spots on the ventral fore- and hindwings of the latter are smaller.

Reversed Roadside-Skipper (*Amblyscirtes reversa*)

The range of this small skipper is from southeast VA, through the Carolinas, extending into northern GA, and then along the Gulf Coast from west FL to MS. It is rarely seen, found in small, local colonies, and, like the Carolina Roadside-Skipper, is more restricted in range than its host plant, switch cane. The preferred habitat is canebrakes at the edges of blackwater streams and rivers.

ventral

dorsal

Throughout much of its range, there are three broods: late Apr to May, late June to July, and late Aug through early Sept.

Although I've looked for this little roadside-skipper along the Gulf Coast, I've not yet found it. Mary Ann Friedman has verbally reported to me that she has seen it in the western FL Panhandle. She has, by e-mail, described this skipper as flying with Arogos Skippers in wet habitat with abundant cane. Even in the most likely conditions, she indicated, it was difficult to find. It has also been found in Baldwin County in southeastern AL. Roever verbally advised that he found it at the Clark Creek Nature Area in MS, less than a mile from the WFE parish line, in mid-Aug 1984. It has also been reported in Hancock County, MS, in June and Aug as well as in Harrison County at the Big Biloxi Rec Area in Apr and Aug, both locations along the Gulf Coast. These sightings are reflected as an isolated range extension in the Brock and Kaufman (2003) guide.

Bell's Roadside-Skipper (*Amblyscirtes belli*)

This roadside-skipper is uncommon to common, with two flights from Mar to Sept, throughout AR, but is absent in the southern Delta (southeastern) region. It has been listed as a member of the butterfly fauna at Red Slough WMA in east-southeast OK. It has been found along the edges of dense woodlands and moist clearings and fields, at several locations in the piney woods region of east TX along the northwest LA state line. These sightings would suggest it might be found along the northwestern LA border with AR and TX.

I have found it a few times in southwest AR at Rick Evans/Grandview Prairie WMA in late Mar and again in June 2012. It was flying in moist woods, near a small creek with multiple kinds of grasses growing in the understory. It was seen at wild garlic when not basking in slivers of sunlight on low-growing plants in the understory. Unlike Pepper and Salt Roadside-Skippers, it does not assume a "jet fighter" or spread-wing position when perched, but holds its wings straight over its back like a sulphur. This skipper's preferred habitat is hardwood forest bordering rivers and streams where its host plant, Indian woodoats, grows along the bank. River oats have also been identified as a host plant in TN. Its first brood appears in the spring, flying into Aug. There may be as many as three broods in the southern portion of its range.

Other species suggested to me that might possibly turn up in LA include the Hickory Hairstreak (*Satyrium caryaevorum*) and Linda's Roadside-Skipper (*Amblyscirtes linda*).

QUESTIONABLE LOUISIANA RECORDS
Dina Yellow (*Eurema dina*)

While researching this book, I discovered that six Dina Yellows were reported as part of the 2002 Lower Pearl River count. As I consider this to be a very unusual sighting, I thought it odd that it wasn't noted as new to the count or mentioned in the field notes. This sulphur is typically (in the broad sense) found in extreme south FL and the Rio Grande Valley of TX. There are no records I could find of it straying north of those immediate areas. As such, this report is either a misidentification or a mistake made while completing the count forms.

ventral

dorsal male

Julia Heliconian (*Dryas iulia*)

I've seen this longwing in both south FL and south TX, but not in LA (or anywhere close). This tropical species has been known to occur sporadically in the Houston area, with all sightings from Sept through Nov. These records include sights from as far east as Baytown. In southeast TX and south FL, the host plant is reported to be vines in the passionflower family, including the native maypop, which is sold and grown as an ornamental throughout LA.

As outlined in my discussion of Barred Yellows, I have questioned three of Lambremont's (1954) listings based on old, single records, all from LAFA. Specifically, I would have expected any record(s) for this species to be farther south and west, such as in CAM. Rather, the only record has been identified as a male captured on Nov 5, 1941, at the city of Lafayette. To date, I have been unsuccessful in my efforts to locate the collection in which this specimen was reported to be included (sent to Lambremont by Dr. Lewis Graham of Southwestern Louisiana Institute), but it is clear that Lambremont actually examined that collection. In his master's thesis, Lambremont referred to the single reported specimen as a straggler and a northern- and eastern-range extension. I continue to wonder if this and the record of a single Great Spangled Fritillary might have been a mistake, but, without the collection, I only question, rather than reject, Lambremont's inclusion of this butterfly within LA's fauna.

Great Spangled Fritillary (*Speyeria cybele*)

This butterfly is regularly found in southwest AR, including at Rick Evans/ Grandview Prairie WMA in Hempstead County. It has also been reported in Lincoln County in southeast AR. This butterfly is also present in north MS, and, in fact, I have seen it in Chickasaw County in May. Thus a sighting from north LA would not be unexpected; however, the only report of this fritillary for LA is of a male found in the city of Lafayette on Oct 2, 1931. Again, this specimen

was part of a collection sent to Lambremont by Dr. Lewis Graham from Southwestern Louisiana Institute. Lambremont recognized that this fritillary is not common in LA (I would describe it as exceedingly rare), and thought it was a straggler that had migrated south during cooler months. Given that the female Florida White I caught in LAFA in 2001 was completely unexpected, as I would no more expect to see a Florida White in LAFA than a Julia or a Great Spangled Fritillary, I guess anything is possible.

Persius Duskywing (*Erynnis persius*)

Although this butterfly is listed from LA on BAMONA (allegedly seen in the Lake Charles area in southwest LA), I have been unable to find any record of when, where, or by whom it was reported. Sources such as Glassberg, Brock and Kaufman, Cech and Tudor, and Scott list it from the Northeast, upper Midwest, and the far West. The closest those sources show it to LA is in east TN and west NC, in the Appalachians.

Leonard's Skipper (*Hesperia leonardus*)

This large skipper has a range primarily north of LA but does extend irregularly into central AR and southeastern OK. It is described as a late-fall skipper with only one brood, Sept into mid-Oct. Herschel Raney had no records closer than the center of AR. In the Bess Report, this skipper was reported as rare at both the Dry Lost Creek and Lake Glades units in Saline County and the Miller County Sandhills Preserve. No specific dates were given for any of those locations. It has also been reported from Red Slough in extreme southeastern OK.

There is some confusion associated with its presence in LA. Lambremont's master's thesis (1951) references an Aug 20, 1947, record from "uptown" New Orleans, not a section of the state where I would anticipate it might be recorded. Then, three years later, in his 1954 study, referenced throughout this project, Lambremont made no reference to the ORL record, but reported only a single specimen from MAD, caught between 1926 and 1931 by Glick in the Tallulah area, in an airplane trap at an altitude of twenty feet. These traps were basically two screen-covered boxed frames hung on each side of a biplane's fuselage, under the top wing. Through the kind efforts of Ann Case with the Howard-Tilton Memorial Library at Tulane, I was able to get a copy of Glick's 1939 study. It confirms what Lambremont reported but doesn't give an exact date of collection. Further, Glick commented, "most of the insects were completely smashed by coming in contact with the screens at high speeds."

I have wondered if that specimen (possibly a female Sachem) might have been misidentified. If this skipper has been present in the past in southwestern AR, it could have also strayed into north LA. Its preferred habitat is varied, but open areas with its host plants, little bluestem and gama grass, are preferred. Males are reported to both perch and patrol. It is attracted to plants with tall flower stems like thistle, on which I have found it in south-central OK.

dorsal

Common Streaky-Skipper (*Celotes nessus*)

I've seen this skipper only in TX in the Austin and San Antonio areas. It is a small skipper with habits much like the checkered-skippers, flight low to the ground, perching often in the grass. The only record from LA of which I am aware was reported by Lambremont (1954), who noted one specimen was collected by Glick (1939), "who did extensive work with traps carried by airplanes," between 1926 and 1931. Glick reported taking one specimen in MAD at an altitude of twenty feet. I have referenced Glick's article within the account for Leonard's Skipper, and I find it odd that he would capture two skippers in MAD (this one and Leonard's Skipper) so far from their typical range. In TX, it reportedly uses members of the mallow and sida families as its host plants.

Dotted Skipper (*Hesperia attalus*)

There are two separate populations of Dotted Skippers, along the East Coast south into FL and the Gulf Coast and then in the plains of KA, OK, and north TX. The historical range of the eastern population included along the Gulf Coast in south MS and extreme southeast LA (Opler and Malikul, 1984; Glassberg, 1999). There are two old records from Pearl River County and Harrison County, MS, both in Sept. Although it is listed on BAMONA as part of LA's fauna, I've found no independent record of this butterfly in LA. The only BAMONA record was from STA, east of Abita Springs, with no other data. Past Lower Pearl River counts reflect that three Dotted Skippers were reported as seen during the 1976 count. Unfortunately, the current count complier, Auld, was not involved in that count before 1993, and had no additional information about the reported sightings.

Mary Ann Friedman, who resides in the Pensacola, FL, area, was kind enough to send via e-mail the pictures included in this book. She reported finding this skipper in the extreme western section of the FL Panhandle, flying with Arogos Skippers. Her impressions were that the two species often flew at the same time. Friedman described the habitat as pine flatwoods, including both wet pitcher-plant-type areas and more upland, drier habitats consistent with the pine savannas of STA, so it is not inconceivable that it was once, and might again be, present in the FL Parishes.

In FL, it is reported to have flights from Mar to May and then from July to Nov, which may reflect more than two broods. Males are reported to be territorial, and both sexes visit flowers from mid-morning to mid-afternoon. They may become inactive after one o'clock on hot days. The food plant in FL was identified as wiregrass. It may also use bluestems or lopsided Indian grass. Friedman found it in areas with grasses like false rosemary (Apr) and coastal-plain honeycomb-head (Oct).

ventral male

dorsal male

ventral female

APPENDIX B

Common and Scientific Names of Referenced Plants

acacia (*Acacia* ssp.)
acanthus (*Acanthaceae* ssp.)
Adam's Needle (*Yucca smalliana*)
alfalfa (*Medicago sativa*)
althea (*Hibiscus syriacus*)
American bladder nut (*Staphylea trifolia*)
American elm (*Ulmus americana*)
American glasswort (*Salicornia virginica*)
American hogpeanut (*Amphicarpaea bracteata*)
American wild carrot (*Daucus pusillus*)
Apalachicola wild indigo (*Baptisia megacarpa*)
aquatic milkweed (*Asclepias perennis*)
Argentine senna (*Senna corymbosa*)
arrowfeather threeawn (*Aristida purpurascens*)
ash (*Fraxinus* ssp.)
asters (*Aster* ssp.)
Atlantic white cedar (*Chamaecyparis thyoides*)
bamboo or cane (*Arundinaria* ssp.)
basswood (*Tilia americana*)
bear grass (*Yucca filamentosa*)
bearded skeletongrass (*Gymnopogon ambiguus*)
beebalm (*Monarda* ssp.)
beggar's ticks aka tick trefoil or tick clover (*Desmodium* ssp.)
bent grass (*Agrostis* ssp.)
Bermuda grass (*Cynodon dactylon*)
big bluestem (*Andropogon gerardii gerardii*)
bilberry (*Amelanchier arborea*)

bittercress (*Cardamine hirsuta*)

black cherry (*Prunus serotina*)

black locust (*Robinia psuedoacaia*)

black titi (*Cliftonia monophylla*)

black walnut (*Juglans nigra*)

black willow (*Salix nigra*)

blackberry (*Rubus* ssp.)

black-eyed-Susans (*Rudbeckia* ssp.)

blue porterweed (*Strachytarphata jamaicensis*)

blueberries (*Vaccinium* ssp.)

bluegrass (*Poa* ssp.)

branching foldwing (*Dicliptera branchiata*)

Brazilian plume (*Jacobinia carnea*)

Brazilian vervain (*Verbena brasiliensis*)

broomsedge (*Andropogon virginicus*)

bull thistle (*Cirsium vulgare*)

butterfly bush (*Buddleia* ssp.)

butterfly pea (*Clitoria mariana*)

butterfly weed (*Asclepias tuberosa*)

butterworts (*Pinguicula* ssp.)

buttonbush (*Cephalanthus occidentals*)

cabbage (*Brassica* ssp.)

candelabra plant (*Cassia alata*)

cannas (*Cannaceae* ssp.)

Carolina vetch (*Vicia caroliniana*)

centipede grass (*Eremochloa ophiuroides*)

citrus trees (*Citrus* ssp.)

chestnut (*Quercus montana*)

clover (*Trifolium* ssp.)

cockscomb (*Celosia* ssp.)

coffee senna (*Cassia occidentalis*)

coffeeweed (*Sesbania exaltata*)

coral bean (*Erythrina herbacea*)

crabgrass (*Digitaria* ssp.)

cream wild indigo (*Baptisia bracteata*)

crotons (*Croton* ssp.)

cudweed (*Gnaphalium* ssp.)

dallisgrass (*Paspalum dilatatum*)

dahoon (*Ilex cassine*)

deertongue (*Panicum clandestinum*)

devil's walkingstick (*Aralia spinosa*)

dill (*Anethum graveolens*)

dogweed (*Thymophylla* ssp.)

downy false foxglove (*Aureolaria virginica*)

Dutchman's pipe (*Aristolochia macrophylla*)

Eastern gamma (sometimes spelled as gama grass) (*Tripsacum dactyloides*)

Eastern red cedar (*Juniperus virginiana*)

elm (*Ulmus* ssp.)

English plantain (*Plantago lanceolata*)

fall witchgrass (*Leptoloma cognatum*)

false foxgloves (*Aureolaria* ssp. including *Aureularia flava*)

false indigo (*Amorpha fruticosa*)

fennel (*Foeniculum vulgare*)

field grass (*Andropogon* ssp. including *Andropogon virginicus*)

firespike (*Odontonema strictum*)

flat sedge (*Cyperus* ssp.)

Florida hairy wild indigo (*Baptisia calycosa* var. *villosa*)

frog fruit aka carpetweed (*Lippia* ssp.)

giant cane (*Arundinaria gigantean*)

giant cutgrass aka water or marsh millet (*Zizaniopsis miliacea*)

giant or tall sunflower (*Helianthus gigantus*)

goat's rue (*Tephrosia virginiana*)

goldenrod (*Solidago* ssp.)

great ragweed (*Ambrosia trifida*)

green beans (*Dolichos lablab*)

hackberry (*Celtis laevigata*)

Hercules club (*Zanthoxylum clava-herculis*)

hickory (*Carya* ssp.)

honey mesquite (*Prosopis glandulosa*)

huckleberry (*Gaylussacia*)

hyacinth bean (*Dolichos lablab*)

Indian grass (*Sorghastrum nutans*)

Indian shot (*Canna indica*)

Indian sea oats / Indian woodoats (*Chasmanthium latifolium*) ironweed (*Vernonia* ssp.)

ixora (*Ixora coccinea*)

Johnson grass (*Sorghum halepense*)

King's crown (*J. suberecta*)

lamb's quarters (*Chenopodium album*)

lance-leaved waterwillow aka looseflower waterwillow (*Justicia ovata*)

lantana (*Lantana* ssp.)

leatherwood (*Cyrilla racemiflora*)

legumes (*Leguminosae* ssp.)

liatris (*Liatris* ssp.)

lignum vitae (*Guaiacum* ssp.)

little bluestem (*Andropogon scoparius*)

locust (*Robinia* ssp.)

longleaf woodoats (*C. sessiliflorum*)

lopsided Indian grass (*Sorghastrum secundum*)

maidencane (*P. hemitomon*)

mallows (*Malvaceae* ssp.)

mannagrass (*Glyceria* ssp.)

Mexican milkweed (*Aureularia curassavica*)

milkweeds (*Asclepias* ssp.)

mimosa (*Albizia julibrissin*)

mistletoe (*Phoradendron serotinum*)

mountain mints (*Pycnanthemum* ssp.)

mustards (*Cruciferae* ssp.)

nettles (*Urtica* ssp.)

New Jersey tea (*Ceanothus americanus*)

noseburn (*Tragia* ssp.)

oaks (*Quercus* ssp.)

one-flower honeycomb-head (*Balduina uniflora*)

panicgrass (*Panicum* ssp.)

parsley (*Petroselium crispum*)

partridge pea (*Chamaecrista cinerea* and *C. fasiculata*)

passion vine/passionflower (*Passiflora*, particularly *P. incarnata* and *P. caerulea*)

pawpaws (*Asimina* ssp.)

pea family (*Fabaceae* ssp.)

Pennsylvania bittercress (*Cardamine pensylvanica*)

peppergrass (*Lepidium* ssp.)

perennial glasswort (*Sarcocornia ambigua*)
pickerelweed (*Pontederia cordata*)
pigweed (*Amaranthaceae* ssp.)
pink glove (*Agalinis fasciculata*)
pipevine (*Aristolochia* ssp.)
plum (*Prunus angustifolia*)
plumbago aka leadwort (*Plumbago auriculata*)
poplars (*Populus* ssp.)
potato bean (*Apios americana*)
privat aka ligustrum (*Ligustrum sinense*)
purple coneflower (*Echinacea purpurea*)
purple passion-vine aka maypop (*Passiflora incarnata*)
purpletop tridens (*Tridens flavus*)
pussytoes (*Antennaria parlini*)
Queen Anne's lace (*Daucus carota*)
ragweed (*Ambrosia* ssp.)
rattlebox (*Crotalaria* ssp.)
red bay (*Persea borbonia*)
red buckeye (*Aesculus pavia*)
red clover (*Trifolium pratense*)
red pentas (*Pentas lanceolata*)
redbud (*Cercis canadensis*)
redtop panicum (*Panicum rigidulum*)
reeds (*Phragmites* ssp.)
rice (*Oryza sativa*)
river oats (*Chasmanthium latifolium* or *Unifola latifolia*)
rockcress (*Arabis* ssp.)
roundleaf bacopa (*Bacopa rotundifolia*)
roundleaf ragwort (*Senecio obovatus*)
ruellia (*Ruellia* ssp.)
rushes (*Scirpus* ssp.)
saltgrass (*Distichlis spicata*)
salt-marsh bulrush (*Schoenoplectus [Scirpus] robustus*)
saltwort (*Batis maritima*)
sassafras (*Sassafras albidum*)
savannah panicgrass (*Panicum gymnocarpon*)
sawgrass (*Cladium jamaicense*)

sedges (*Carex* ssp.)
senna (*Senna* ssp.)
shining sumac (*Rhus copallina*)
shoreline sedge (*Carex hyalinolepis*)
shrimp plant (*Beloperone guttata*)
sickle-pod (*Cassia obtusafolia*)
sida (*Sida* ssp.)
sideoats (*Bouteloua curtipendula*)
slender woodoats (*Chasmanthium laxum*)
small rockcresses (*Brassicaceae* ssp.)
smooth false foxglove (*Aureolaria laevigata*)
snapdragons (*Antirrhinum* ssp.)
sneezeweed (*Helenium* ssp.)
southern sea-blite (*Suaeda linearis*)
Spanish bayonet (*Yucca aloifolia*)
Spanish dagger (*Yucca gloriosa*)
spicebush (*Lindera benzoin*)
St. Augustine grass (*Stenotaphrum secundatum*)
sumac (*Rhus* ssp.)
sundews (*Drosera* ssp.)
sunflowers (*Helianthus* ssp.)
swamp doghopple (*Leucothoe racemosa*)
swamp laurel (*Magnolia virginiana*)
swamp thistle (*Cirsium muticum*)
swampbay (*Persea palustris*)
sweet acacia (*Acacia farnesiana*)
sweetleaf (*Symplocus tinctoria*)
switch cane (*Arundinaria tecta*)
switchgrass (*Panicum virgatum*)
tall thistle (*Circium altissiumum*)
thicket aka wild bean (*Phaseolus polystachios*)
tickseed (*Coreopsis* ssp.)
toadflax (*Lineria* ssp.)
toothache grass (*Ctenium aromaticum*)
tumble-grass (*Eragrostis spectabilis*)
vetches (*Vicia* ssp.)
viburnum (*Viburnum* ssp.)

wax myrtle (*Myrica cerifera*)
weedy plantains (*Plantago* ssp.)
white clover (*Trifolium repens*)
wild indigo (*Baptisia* ssp.)
willow (*Salix* ssp.)
winged pitcher plants (*Sarracenia alata*)
wingstem crownbeard (*Verbesina walteri*)
winter cassia (*Cassia bicapsularis*)
wiregrass (*Aristida stricta* var. *beyrichiana*)
wisteria (*Wisteria* ssp.)
witch hazel (*Hamamelis virginiana*)
woodland sunflower (*Helianthus divaricatus*)
woolgrass (*Scirpus cyperinus*)
woolly beardgrass (*Erianthus divaricatus*)
woolly croton (*Croton capitatus*)
yaupon (*Ilex vomitoria*)
yellow false foxglove (*Aureolaria grandiflora*)
yellow passionflower (*Passiflora lutea*)
yellow thistle (*Cirsium horridulum*)
yellow wild indigo (*Baptisia tinctoria*)

GLOSSARY

abdomen: the last body part behind the head and thorax.

aberration: an abnormal individual. In the context of butterflies, one that is not normal in appearance for that species.

alba: white form.

anal spots: spots at tail end or anal angle of wing.

antenna: one of two long, slender appendages extending from the head, used for smelling.

aposematic: refers to coloring that presents a warning of distastefulness.

bait trap: a type of insect trap, of varying designs, which uses a bait such as rotting bananas, other types of fruit, or dung to attract insects into the trap.

basal cell: the wing cell nearest the body.

bask: expose wings and body to the sun to absorb heat energy and thereby raise body temperature.

Batesian mimicry: when an edible species imitates in appearance a toxic species (see **mimicry**).

border: the outer edge of the wings.

brood: each generation of a species; eggs that hatch and mature into adult butterflies at approximately the same time.

burn(s)/burning: also known as controlled or prescribed burning, a technique used in forest management on a rotating basis to control secondary and successional growth.

carnivorous: feeding primarily on other animals/insects.

cells: the space between wing veins.

cheniers: long, narrow beach ridges along the Louisiana Gulf Coast, especially in Cameron Parish, often heavily wooded with oaks as the predominant tree and separated by mudflats, marsh, and/or wet prairie.

cline/clinal intermediate: a gradual geographic blending or separation of characteristics of species or subspecies.

colony: a local group of individuals belonging to the same species, isolated from other colonies by geography.

congeneric: refers to members of the same genus of butterflies.

coniferous: refers to softwoods, typically pines in the southern US.

cryptic: serving to conceal, as in a cryptically green chrysalis.

deciduous: refers to woody trees and plants that drop their leaves during the winter.

dimorphic: possessing two forms, related to sex, season, or other factors.

dorsal: the upper or top side of a wing, above.

dwarf: a size aberration, caused by environmental, developmental, or genetic factors.

extirpated: removed or destroyed from a particular area.

eyespots: rings and/or spots on a butterfly's wings that resemble an eye.

falcate: hooked or curved, as in a forewing tip.

fauna: all the animal species in a defined area.

flight: a single generation of adult butterflies of a certain species that are visibly on the wing at generally the same time.

forewings: the anterior (top) set of wings.

form: a color, geographic, or seasonal variation of a species, not a formal family or species.

fringe: scales that stick out along the edges of the wings.

generalist: a butterfly species that uses a variety of host plants from different plant families.

genitalia: sex organs and their support structures.

glassine: see **hyaline spots.**

habitat: an area in which a certain species of butterfly prefers to live, including the type of plant community needed for survival.

hardwood: a forested region dominated by hardwood trees, primarily deciduous trees.

hindwings: the posterior (bottom) set of wings.

host plant: any plant that a caterpillar will eat or on which eggs are typically laid.

hyaline spots: clear or translucent wing spots that lack scales and are transparent.

hybrid: a sexual cross between two separate but (usually) related species that generates a viable adult.

instar: an individual stage of larval development.

interbreed: to successfully mate in nature and generate healthy offspring between two species or subspecies.

intergrades: offspring that are intermediate in appearance, produced as a result of interbreeding in a geographic area between two subspecies.

irruption: a sudden or rapid and significant increase in an animal population.

labial palpi: the elongated mouthparts that are on either side of the proboscis.

larvae: in butterflies, larvae are commonly called caterpillars.

legume: a plant in the pea family.

local: refers to butterflies limited in distribution to a particular area.

migratory: refers to movement of adult butterflies into a new geographic area to breed or feed, often seasonal.

mimicry: when one butterfly species imitates or copies a model for protection.

Mullerian mimicry: when a toxic species imitates in appearance another toxic species.

nominate: the primary or original subspecies of a species.

oviposit: the act of a butterfly laying eggs.

patrol: mate-locating behavior in which a male flies continuously in search of a mate, sometimes flying a repeated route.

perch: mate-locating behavior in which an adult butterfly (usually male) lands and remains on a stem, leaf, blade of grass, stick, etc., to wait for a passing mate.

pine flats/pine flatlands: pine forests on broad flats that have poorly drained soils, often interspersed with swamps and/or grassy prairies.

pitcher-plant bog: a wet prairie, bog, or seep inhabited by carnivorous pitcher plants as well as wildflowers, typically on the side of or at the base of a hill in pine forests.

phenotypes: the traits of a species that are expressed physically.

pheromone trap: a type of insect trap, of varying designs, which uses pheromones, typically sex pheromones, to attract insects into the trap.

population: all the adults of a particular species that live in sufficient proximity to each other to mate.

proboscis: a butterfly's mouth parts, typically a flexible feeding tube that is coiled when not in use.

puddling: the gathering of butterflies (typically males) at wet spots.

pupa: the third (chrysalis) stage of a butterfly's metamorphosis.

race: a geographically distinct form of a species.

range: the geographic area where a species of butterfly lives, including regular migratory routes.

ray: a partial stripe or streak of different coloring across a wing.

rosa: a form that occurs when the wings are tinged with pink or red.

sandhills: isolated areas of exposed sand and sparse vegetation surrounded by denser forests.

scales: tiny shingle-like structures that cover a butterfly's wings and generate color and patterns as a result of pigmentation and/or a prism effect.

seasonal forms: forms that are produced based upon the seasons.

secondary growth: successional plant growth that develops after clearing or burning of the original forest.

seep: a moist or wet place where groundwater reaches the surface, typically in nonflowing puddles.

sibling species: two or more species that closely resemble each other but cannot successfully interbreed.

slough: a shallow channel sporadically filled with slow-moving or stagnant water.

species: a group of butterflies with common attributes and capable of natural reproduction; the basic unit for classifying living organisms.

specimen: a sample of something (such as a species) intended to represent a larger population.

stand: a contiguous community of trees or grass of sufficient uniformity of composition so as to be distinguishable from other plant communities.

stray: a butterfly that rarely or infrequently moves into a region it is known to inhabit, not to be expected on a recurring basis.

subspecies: a subgrouping of a species that occupies a distinct geographic range and that differs in appearance from other subspecies of that same species.

succession: the process of more or less predictable species turnover over time in a particular plant community.

symbiotic: refers to a close, interdependent association between two or more different species, to the benefit of one or both species.

sympatric: refers to different species flying in an overlapping range.

synchronic: refers to species, subspecies, or forms flying at the same time.

tail: an extension at the anal angle of the hindwing in some butterflies (swallowtails and hairstreaks).

territory: the association of a butterfly with a particular location; an area or space claimed by an individual butterfly and protected by that butterfly from intruders.

toxin: poisonous substance produced within a living organism.

toxicity: degree of a toxin needed to harm another organism.

tropical: refers to butterflies that typically live and breed in a hot, equator-like climate.

ultraviolet: beyond the rays of the violet end of the visible light spectrum, not visible to people but visible to butterflies.

understory: the underlying layer of vegetation below a forest's canopy.

unpalatable: distasteful, not edible due to the presence of toxins.

UV/MV light trap: a type of insect trap, of varying designs, which uses either an ultraviolet light (blacklight) or a mercury-vapor light to attract insects into the trap, typically used at night to collect moths but known to attract butterflies as well.

veins/veining: the hollow and visible structural supports of a butterfly's wings arranged, by species, in a predictable fashion.

ventral: refers to the underside of a butterfly's wings, underneath.

A NOTE ON SOURCES

My first step toward generating an updated list of all known butterflies and skippers in Louisiana was to reference the Butterflies and Moths of North America (BAMONA) website (www.butterfliesandmoths.org). The BAMONA project, hosted by the Butterfly and Moth Information Network, is an "ambitious effort to collect and provide access to quality-controlled data about butterflies and moths for the continent of North America." It brings together in one location "verified occurrence and life history data." The database is a composite of historical data as well as recent and current submissions by "citizen scientists." It includes data searchable by species, state, and even county or parish.

I used the LA list on that website to start the process of identifying the butterflies and skippers that have been recorded within the state. I then turned to a number of other sources aside from my own records to add other species and/or sightings that had been reported but were not included on the BAMONA site. As part of this process, I created a list of what I thought were potentially erroneous, questionable, or dubious records and moved some of the BAMONA records to that list. I also created a list of species that I felt might eventually turn up in LA. Ironically, while I was working on this project, I was required to move one butterfly, the Bordered Patch, from the "possible" list to the recorded list based on a first-time sighting in the fall of 2012.

As is reflected in my References section, I used numerous sources as I attempted to identify what butterflies have been seen and where. My primary reference was Edward Lambremont's 1954 study, *The Butterflies and Skippers of Louisiana*. That source was supplemented by Gary N. Ross and Lambremont's 1963 "Annotated Supplement to the State List of Louisiana Butterflies and Skippers" and Lambremont and Ross's 1965 "New State Records and Annotated Field Data for Louisiana Butterflies and Skippers." I also had access to articles by Bryant Mather ("Louisiana Butterfly Records," covering CON and CAT) and R. O. Kendall ("New Skipper and Butterfly Records for Southwest Louisiana," covering VER, CALC, and BEAU), both from that same general time frame.

Lambremont (1954) referenced several collections he reviewed, including one at Southwestern Louisiana Institute, now the University of Louisiana at Lafayette (ULL). Lambremont reported the presence of several species within the state based on that collection. Because I had questions about three of the reports taken from that collection, I made an effort to locate and view the collection through ULL. Ultimately, after inquiries to the ULL Library and Biology Department, I was advised that no one currently knows the whereabouts of that collection.

Another major source of data was an unpublished 1972 manuscript by the late Gayle T. Strickland, provided to me by the late Bryant Mather (who also sent me the articles referenced above). Through the help of Killian Roever, I was able to make contact with Gayle and supplement the content of his manuscript through access to his collection and his field notes along with verbal communications with him. His field notes reflected his travels around the state during the years 1967–69, visiting ASS, CAD, CAM, CAT, DES, EBR, GRA, IBV, LAFO, LIV, NAT, PCP, SAB, SHE, SJB, WBR, and WFE. The notes contain information about what species he saw each day, the numbers seen, specific locations visited, and other observations. He and his wife created a website that addresses the complexities presented by the Pygmy Blues they had found in LA.

Several people were kind enough to provide their records from around the state. Killian Roever, both verbally and via e-mail, provided a list of the butterflies and skippers he had caught in the Kisatchie National Forest as well as other locations during the early 1970s. I had access to Kreg Ellzey's "life list," which I printed from birdersdiary.com in 2005, that contained numerous records from LA, including multiple records from both NAT and SAB. Personal websites referenced included Frank Dutton's "Butterflies of the Toledo Bend Lake Area" and the late Ronnie Gaubert's "Butterflies of Louisiana" photo gallery (which appears to have expired and is no longer available online).

Both Gary Ross and Vernon Brou were an invaluable source of information through verbal and e-mail exchanges as well as the multiple articles they have written about their LA experiences and records. The References section of this book includes twenty-six articles and books authored or co-authored by Ross and eighteen articles authored or co-authored by Brou. Several of Brou's studies provided detailed records on LA's hairstreaks, as well as other butterflies. Brou also allowed me to access his large collection. He has been collecting moths, butterflies, and skippers for over forty years, primarily in SJB and STA at his places of residence. The vast majority of the butterflies and skippers he has collected over the years were via bait traps (using several of his original recipes), pheromone traps, and MV light traps. He reported that he averaged collecting thirty to forty thousand butterflies each year for the past four decades using these methods. On occasion, he collected as many as four hundred butterflies of a single species on a single night in his UV/MV light traps. These experiences are reflected in a unique article published in 1974, "Butterflies Taken in Light Traps."

With the help of Victoria Bayless and Mike Ferro, both with the Department of Entomology at Louisiana State University, I gained access to the butterfly and skipper collection maintained at the Louisiana State Arthropod Museum in Baton Rouge. To my surprise, this source generated numerous records I had not previously found. I also picked up additional data concerning records generated by Lambremont, Ross, Strickland, Michael Israel, and Brou.

Several parishes within the state have fewer than twenty-five species reported, and two of those parishes (Lincoln and Ouachita) are also the home of major universities. With the help of members of the biology departments at both schools (Natalie Clay at Louisiana Tech University and Dennis Bell at the University of Louisiana at Monroe), I was able to obtain extensive additional data for both of those parishes as well as surrounding parishes. Jeff Trahan inspected the schools' collections and provided the names of those species within each collection, where the specimens were collected (by parish), and collection dates for inclusion in this project. Both collections significantly increased the amount of data for north-central and northeastern LA.

Jeff Trahan and Rosemary Seidler provided a large volume of data on the northwestern portion of the state. Seidler's LA butterfly listserv was a good source of recent sightings. Trahan's data included an article detailing records from CAD, "The Butterflies of Caddo Parish, Louisiana," as well as a webpage that identified some specific locations within that parish. In 2017, Trahan created a second website, "Butterflies of Louisiana," which I also referenced. Both Trahan and Seidler shared their own records with dates and places from surrounding parishes as well as other portions of the state. And Jeff was instrumental in the completion of this project through his assistance in converting my rough parish range maps into a publishable format and by editing several chapters. Dave Patton provided numerous records from around the state, with emphasis on CAM, LAFA, EVA, VER, SMA, and PCP. Linda Auld provided, by e-mail, her personal experiences with several species, as well as spreadsheets for all years of the Lower Pearl River, Metro New Orleans, and Bonnet Carré counts and BugStock (a weekend-long entomology, natural science, and music event) in Washington, LA. Linda also contributed a study of the Pearl River Basin.

I utilized three studies that reported the results of a three-year study on the Cajun Prairie and southwestern regions of LA. The first was "Butterflies of the Cajun Prairie," by Charles Allen and Malcolm Vidrine. The second, by Vidrine, Charles M. Allen, Bruno Borsari, and Larry Allain, was *Lepidopteran and Odonate Communities in the Cajun Prairie Ecosystem in Southwestern Louisiana*. The third, "Flight Records of Papilionoidea in Southwestern Louisiana," was by Vidrine, Allen, and H. D. Guillory. They defined the study area as the prairie portion of southwest LA isolated by the Big Thicket in TX on the west and the Atchafalaya River to the east, consisting of all or portions of western SLA, western LAFA, ACA, JFD, northern VER, southern EVA, southeastern ALL, and eastern CALC. Other parishes covered include AVO, BEAU, CAM, EVA, IBE, LAFA, RAP, SAB, SLA, and VER. While none of the studies covered skippers in that region, they provided a wealth of information on the butterflies found there. Both Charles Allen and Malcolm Vidrine graciously responded to my questions, and Malcolm provided me with his daily diaries, which contained the raw data on which their studies were based. In fact, the raw data included some skipper material that I was able to incorporate.

I used Season Summary records gathered by the Lepidoptera Society. I had my own copies of the Season Summaries from 1996 forward and made use of the Lepidoptera Society's online database. These records were particularly helpful to supplement my data in the FL Parishes and the lower parishes, as they contained numerous reports by Michael Israel for the former area and Kevin Cunningham in the latter. Kevin and Michael Lefort were also kind enough to share their data on the lower southeastern parishes. Jonathan Clark provided data for some of the lesser-reported central parishes. Jeff McMillian shared his sightings (supported by photographs) associated with his business, Almost Eden Nursery in Merryville, Beauregard Parish. Loice Kendrick-Lacy, master gardener, author of *Gardening to Attract Butterflies*, and director of the annual Haynesville Celebration of Butterflies, also provided a list of butterflies she had seen in and around Haynesville in Claiborne Parish, with associated comments.

During the process of researching this book, I learned that Michael Israel had written his doctoral dissertation on the butterflies of Tunica Hills. Michael was kind enough to not only provide me with a copy of his species accounts from that thesis, but also to send a catalog of the butterflies of Asphodel Plantation, which he had prepared in pamphlet form. These two sources provided a great amount of information about EFE and WFE. The dissertation presented data from his surveys of WFE in the years 1977–80. The pamphlet provided his records from his home near Jackson, LA, through 2007.

Finally, I have made reference to all of the LA Fourth of July count results since 1991, maintained and published by the North American Butterfly Association. These counts were particularly helpful to identify what butterflies and skippers have been seen in the eastern and southeastern regions of the state, where several counts have been held for many years. The Cameron count results also served to supplement records from that parish by David Patton, Rosemary Seidler, and myself.

I have had the pleasure of handling five counts in LA—Allen Acres, Catahoula NWR, Copenhagen Hills, Indian Bayou, and Thistlethwaite. The latter two were started by Gary Ross, and I must again thank him for showing me the ropes and then allowing me to assume responsibility for each. I have also participated in all of the Kisatchie/NAT counts as well as one of the Cameron and Shreveport counts and two Alexandria counts, and have used those records as part of this effort.

My own records are based on my travels around the state since 1995. They are primarily from locations west of the Mississippi River, but also include the Felicianas, TAN, WAS, STA, EBR, and LIV. I have incorporated data from numerous articles I have written and published in the Southern Lepidopterist Society newsletter, including "The Butterflies of Acadiana," "The Butterflies and Skippers of CenLa," and "The Butterflies of the Felicianas." It is in these regions, plus CAM, that the bulk of my records were gathered.

REFERENCES

Ajilvsgi, Geyata. 1990. *Butterfly Gardening for the South*. Dallas: Taylor Publishing Co.

Allard, Simone Hébert. 2013. *Manitoba Butterflies: A Field Guide*. Winnipeg, MB: Turnstone Press.

Allen, Charles M., and Malcolm F. Vidrine. 1990. "Butterflies of the Cajun Prairie." *Louisiana Conservationist* 42 (2): 16–21.

Allen, Charles M., Dawn Allen Newman, and Harry H. Winters. 2004. *Grasses of Louisiana*. 3rd ed. Pitkin, LA: Allen's Native Ventures.

Allen, Thomas J., Jim P. Brock, and Jeffrey Glassberg. 2005. *Caterpillars in the Field and Garden: A Field Guide to the Butterfly Caterpillars of North America*. New York: Oxford University Press.

Auld, Linda. 2013. "Counting Butterflies in the Pearl River Basin." Orleans Audubon Society website, www.jjaudubon.net/node/2236.

Beiriger, Robert. 2003. "Life History of *Megathymus Yuccae* on *Yucca Elephantipes* in South Florida." *Southern Lepidopterists' News* 25 (1): 18–20.

Belth, Jeffrey E. 2013. *Butterflies of Indiana: A Field Guide*. Bloomington: Indiana University Press.

Bess, James. 2000. "A Final Report on Insect Surveys at Three Arkansas Natural Area Complexes with Recommendations for Future Inventory, Monitoring and Habitat Management" ("The Bess Report"). Little Rock: Nature Conservancy Arkansas Field Office.

Boothe, Bill. 2009. "Notes on the Backyard Biology of the Harvester Butterfly (*Feniseca Tarquinius [Fabricius]*) from the Florida Panhandle." *Southern Lepidopterists' News* 31 (3): 113–16.

Bosco, R. W., M. C. Minno, and D. M. Wright. 2015. "A Case of Mistaken Identity: The True Host of the Golden-Banded Skipper *Autochton Cellus* (Hesperiidae: Eudaminae) in the Eastern U.S." *News of the Lepidopterists' Society* 57 (2): 56–59.

Brock, Jim P., and Kenn Kaufman. 2003. *Kaufman Field Guide to Butterflies of North America*. New York: Houghton Mifflin.

Brou, Vernon Antoine, Jr. 1974. "Butterflies Taken in Light Traps." *Journal of the Lepidopterists' Society* 28 (4): 331.

———. 1993. "Voltinism of *Cercyonis Pegala Abbotti* F. M. Brown." *Southern Lepidopterists' News* 15 (3): 27–28.

———. 2000. "*Parrhasius M-Album* (Lycaenidae) in Louisiana." *Southern Lepidopterists' News* 22 (2): 31.

———. 2005. "*Calycopis Cecrops* (F.) in Louisiana." *Southern Lepidopterists' News* 27 (1): 5–6.

———. 2007a. "*Phoebis Sennae Eubale* Linnaeus in Louisiana." *Southern Lepidopterists' News* 29 (2): 33–34.

———. 2007b. "*Satyrium Liparops* (Leconte, 1833) in Louisiana." *Southern Lepidopterists' News* 29 (1): 19–20.

———. 2007c. "*Satyrium Ontario* (W. H. Edwards) in Louisiana." *Southern Lepidopterists' News* 29 (1): 17–18.

———. 2007d. "Spotlight on Rearing: *Papilio Palamedes* Drury in Louisiana." *Southern Lepidopterists' News* 29 (1): 13–15.

———. 2008a. "*Nymphalis Milberti* (Godart) (Lepidoptera: Nymphalidae) Captured in Louisiana." *Southern Lepidopterists' News* 30 (3): 109–10.

———. 2008b. "*Satyrium Kingi* (Klots and Clench) in Louisiana." *Southern Lepidopterists' News* 30 (1): 18–19.

———. 2009a. "*Satyrium Calanus Falacer* (Godart [1824]) (Lepidoptera: Lycaenidae) in Louisiana." *Southern Lepidopterists' News* 31 (1): 1–2.

———. 2009b. "*Utetheisa Ornatrix* (L., 1758) and *Utetheisa Bella* (L., 1758) (Lepidoptera: Arctiidae) in Louisiana." *Southern Lepidopterists' News* 31 (3): 119–23.

———. 2013a. "*Anaea Andria* Scudder (1875) (Lepidoptera: Nymphalidae) in Louisiana." *Southern Lepidopterists' News* 35 (2): 86–87.

———. 2013b. "Earliest Records for *Leptotes Cassius* (Cramer, 1775) in Louisiana." *Southern Lepidopterists' News* 35 (3): 123.

———. 2013c. "The *Enodia* Hübner (Lepidoptera: Nymphalidae) of Louisiana." *Southern Lepidopterists' News* 35 (1): 29–30.

———. 2015a. "*Polygonia Comma* (Harris 1842) (Lepidoptera: Nymphalidae) in Louisiana." *Southern Lepidopterists' News* 37 (1): 19–20.

———. 2015b. "*Protographium Marcellus* (Cramer) (Lepidoptera: Papilionidae) in Louisiana." *Southern Lepidopterists' News* 37 (1): 25–26.

Brou, Vernon Antoine, Jr., Michael T. Lefort, and Kevin J. Cunningham. 2008. "*Anartia Jatrophae Guantanamo* Munroe in Louisiana." *Southern Lepidopterists' News* 30 (3): 100–101.

Burns, J., 2000. "*Pyrgus Communis* and *Pyrgus Albescens* Are Separate Transcontinental Species with Variable but Diagnostic Valves." *Journal of the Lepidopterists' Society* 54 (2): 52–71.

Butterflies and Moths of North America website. www.butterfliesandmoths.org.

Calhoun, John V. 2004. "First Confirmed Record of *Erynnis Martialis* (Scudder) in Florida." *News of the Lepidopterists' Society* 46 (3): 78, 77.

———. 2017. "A Reevaluation of *Papilio Pegala* and *Papilio Alope* F. with a Lectotype Designation and a Review of *Cercyonis Pegala* (Nymphalidae: Satyrinae) in Eastern North America." *Journal of the Lepidopterists' Society* 70 (1): 20–46.

Cech, Rick, and Guy Tudor. 2005. *Butterflies of the East Coast: An Observer's Guide.* Princeton: Princeton University Press.

Cong, Qian, and Nick V. Grishin. 2014. "A New *Hermeuptychia* (Lepidoptera, Nymphalidae, Satyrinae) Is Sympatric and Synchronic with *H. Sosybius* in Southeast US Coastal Plains, While Another New *Hermeuptychia* Species—not *Hermes*—Inhabits South Texas and Northeast Mexico." *ZooKeys*, 379: 43. DOI: 10.3897/zookeys.379.6394.

Cunningham, Kevin J. 2011. "A New Louisiana State Record: *Problema Byssus Byssus* (W. H. Edwards)." *Southern Lepidopterists' News* 33 (4): 149.

Douglas, Matthew M. 1986. *The Lives of Butterflies.* Ann Arbor: University of Michigan Press.

Durden, C. J. 1990. *Guide to Butterflies of Austin.* Austin: Balconian Naturalists' Group for the Texas Botanical Garden Society.

Dutton, Frank. Toledo Bend Lake Butterfly Guide. www.toledo-bend.us/index.asp?bfly.

Ehrlich, Paul R., and Anne H. Ehrlich. 1961. *How to Know the Butterflies: Illustrated Keys for Determining to Species All Butterflies Found in North America . . .* Dubuque, IA: Wm. C. Brown Co.

Emmitt, Randy. 2005. *Butterflies of the Carolinas and Virginias.* Interactive CD, Metalmark Press.

Ford, E. B. 1957. *Butterflies.* 3rd ed. London: Collins.

Gatrelle, Ronald R. 1999. "Hübner's *Helicta:* The Forgotten *Neonympha.*" *Taxonomic Report of the International Lepidoptera Survey* 1 (8): 1–8.

———. 2000. "Description of a New Subspecies of *Poanes Aaroni* (Hesperioidea: Hesperiinae) from the West Central Gulf Coast of the Southern United States." *Taxonomic Report of the International Lepidoptera Survey* 2 (2): 1–9.

———. 2001. "*Thorybes* Clarification." *International Lepidoptera Survey Newsletter.* 3 (1): 1–2.

Gibbs, Lawrence, and Orley R. Taylor. 1998. "The White Monarch." Monarch Watch Reading Room. Retrieved July 17, 2014, www.monarchwatch.org/read/articles/nivosus.htm.

Glassberg, Jeffrey. 1999. *Butterflies through Binoculars: The East.* New York: Oxford University Press.

———. 2012. *A Swift Guide to Butterflies of North America.* China: Sunstreak Books.

———. 2013. "Go, Get Set, on Your Marks: Cloudywings in the Genus *Thorybes.*" Part 1. *American Butterflies* 21 (2): 8–19.

Glassberg, Jeffrey, Marc C. Minno, and John V. Calhoun. 2000. *Butterflies through Binoculars: A Field, Finding, and Gardening Guide to Butterflies in Florida.* New York: Oxford University Press.

Glick, P. A. 1939. "The Distribution of Insects, Spiders and Mites in the Air." U.S. Dept. of Agriculture, *Technical Bulletin* 673: 1–150.

Hall, Donald W., Jerry F. Butler, and Marc Minno. 2007. "Harvester Butterfly, *Feniseca Tarquinius* (Fabricius) (Insecta: Lepidoptera: Lycaenidae: Melitinae)." University of Florida IFAS Extension Booklet EENY-404, edis.ifas.ufl.edu/pdffiles/in/in72700.pdf.

Harder, David, Dean Jue, and Sally Jue. 2007. "Definitive Destination: Elinor Klapp-Phipps Park, Leon County, Florida." *American Butterflies* 15 (3/4): 4–15.

Harris, Lucien, Jr. 1972. *Butterflies of Georgia*. Norman, OK: University of Oklahoma Press.

Heitzman, J. Richard, and C. F. dos Passos. 1974. "*Lethe Portlandia* (Fabricius) and *L. anthedon* (Clark), Sibling Species, with Descriptions of New Subspecies of the Former (Lepidoptera; Satyridae)." *Transactions of the American Entomological Society* 100: 52–99.

Heitzman, J. Richard, and Joan E. Heitzman. 1996. *Butterflies and Moths of Missouri*. Jefferson City: Missouri Department of Conservation.

Howe, William H. 1975. *The Butterflies of North America*. Garden City, NY: Doubleday and Co.

Howell, W. Mike, and Vitaly Charny. 2010. *Butterflies of Alabama*. New York: Pearson Learning Solutions.

Irwin, Roderick R. 1972. "Further Notes on *Euphyes Dukesi* (Hesperiidae)." *Journal of Research on the Lepidoptera* 10 (2): 185–88.

Israel, Michael Lawrence. 1981. "The Butterflies and Skippers of the Tunica Hills of Southeastern Louisiana and Southwestern Mississippi." PhD diss., Louisiana State University, Baton Rouge.

Israel, Michael L., and Ida M. Sharp. 2007. *Butterfly Garden Guide, Asphodel Plantation*. Jackson, LA: Self-published.

Jue, Dean, and Sally Jue. 2014. "Elinor Klapp-Phipps Park: A Story of Butterflies, Horses and People." *American Butterflies* 22 (1): 44–46.

Jue, Dean, Sally Jue, and David Harder. 2011. "Elinor Klapp-Phipps Park Update: A Success Story for Golden-Banded Skippers and Land Managers." *American Butterflies* 19 (1): 30–38.

Kendall, Roy O. 1963. "New Skipper and Butterfly Records for Southwest Louisiana." *Journal of the Lepidopterists' Society* 17 (1): 21–24.

Klots, Alexander B. 1951. *A Field Guide to the Butterflies of North America, East of the Great Plains*. Boston: Houghton Mifflin Co.

Koehn, Leroy C. 2011a. "*Colias Eurytheme* and *Colias Philodice*: A Progress Report." *Southern Lepidopterists' News* 25 (2): 37–40, Inserts A and B.

———. 2011b. "Season Summary of 2010." *News of the Lepidopterists' Society* (June): 1–162.

———. 2012. "Season Summary of 2011." *News of the Lepidopterists' Society* (Mar.): 1–174.

Lambremont, Edward Nelson. 1951. "The Butterflies and Skippers of Louisiana." PhD diss., Tulane University, New Orleans.

———. 1954. *The Butterflies and Skippers of Louisiana,* Tulane Studies in Zoology. Vol. 1 (10).

Lambremont, Edward N., and Gary N. Ross. 1965. "New State Records and Annotated Field Data for Louisiana Butterflies and Skippers." *Journal of the Lepidopterists' Society.* 19 (1): 47–52.

Lockwood, Michael. 2003. "Reports of State Coordinators: Louisiana." *Southern Lepidopterists' News* 25 (2): 60–61.

Lombardini, J. Barry. 2006. "Snout Butterflies Are Also in Lubbock, Texas." *Southern Lepidopterists' News* 28 (3): 69–70.

———. 2010. "Dusky-Blue Groundstreak [*Calycopis Isobeon* (Butler & H. Druce, 1872)] Versus Red-Banded Hairstreak [*Calycopis Cecrops* (Fabricius, 1793)]." *Southern Lepidopterists' News* 32 (1): 31–32.

Louisiana Department of Wildlife and Fisheries. 2005. *Louisiana Comprehensive Wildlife Conservation Strategy,* Appendix F: Species of Conservation Concern in Louisiana. www.wlf.louisiana.gov/sites/default/files/pdf/page_wildlife/33691-Wildlife%20Action%20Plan%20Details/la_wap_pdf.pdf

Louisiana listserv. www.freelists.org/list/louisianaleps.

Louisiana State Parks website. www.crt.state.la.us/parks/.

Marks, Craig W. 2007. "The Butterflies of Acadiana." *Southern Lepidopterists' News* 29 (4): 140–45.

———. 2008a. "The Butterflies and Skippers of CenLa." *Southern Lepidopterists' News* 30 (1): 25–31.

———. 2008b. "The Butterflies of the Felicianas." *Southern Lepidopterists' News* 30 (2): 64–72.

———. 2008c. "Dianas in Southwest Arkansas at Rick Evans/Grandview Prairie WMA." *Southern Lepidopterists' News* 30 (4): 136–40.

———. 2009a. "Crossing the Perplexing Waters of the Delaware." *Southern Lepidopterists' News* 32 (1): 25–28.

———. 2009b. "My Search for the 'Frosted' Grail." *Southern Lepidopterists' News* 30 (3): 93–97.

———. 2010a. "Chasing Silver and Gold: Scintillant Metalmarks across the SLS Region." *Southern Lepidopterists' News* 32 (3): 133–40.

———. 2010b. "Louisiana Kings and Queens and Other Things." *Association of Tropical Lepidopterist Notes* (Dec. 2010): 1–2.

———. 2010c. "Thistlethwaite WMA, St. Landry Parish, Louisiana, May 2, 2010." *Southern Lepidopterists' News* 32 (2): 60–61.

———. 2011a. "Addendum to the Article: *Euphyes Bayensis* in Louisiana." *Southern Lepidopterists' News* 33 (3): 120.

————. 2011b. "*Euphyes Bayensis* Located in Louisiana." *Southern Lepidopterists' News* 33 (3): 92–95.

————. 2011c. "Twin Sons of Different Mothers." *Southern Lepidopterists' News* 33 (4): 152–55.

————. 2012. "Mississippi Been Berry, Berry Good to Me." *Southern Lepidopterists' News* 34 (1): 4–6.

————. 2013a. "Another Fine Meske." *Southern Lepidopterists' News* 35 (2): 67–70.

————. 2013b. "North and South: A Tale of Two Butterflies." *Southern Lepidopterists' News* 35 (3): 124–27.

————. 2013c. "To Tell the Truth: Will the Real Arogos Skipper Please Stand Up." *Southern Lepidopterists' News* 35 (4): 185–88.

————. 2014a. "Sometimes It's Heaven, Sometimes It's Hell, Sometimes I Don't Even Know." *Southern Lepidopterists' News* 36 (3): 145–47.

————. 2014b. "Who Are You, Who, Who, Who, Who?" *Southern Lepidopterists' News* 36 (4): 169–71.

————. 2015a. "Byssus Skippers in Southwestern Arkansas." *Southern Lepidopterists' News* 37 (1): 31–34.

————. 2015b. "Metamorphosis Brazilian-Style." *Southern Lepidopterists' News* 37 (3): 124–26.

————. 2015c. "September on the Gulf Coast." *Southern Lepidopterists' News* 37 (4): 165–68.

————. 2016a. "Done and Dusted in Louisiana—Or Is It?" *Southern Lepidopterists' News* 38 (3): 3–8.

————. 2016b. "Oklahoma: Sooner and Later." *Southern Lepidopterists' News* 38 (1): 53–57.

Mather, Bryant. 1963. "*Euphyes Dukesi*: A Review of Knowledge of Its Distribution in Time and Space and Its Habitat." *Journal of Research on the Lepidoptera* 2 (2): 161–69.

————. 1966a. "Louisiana Butterfly Records." *Journal of the Lepidopterists' Society* 20 (2): 102.

————. 1966b. "*Speyeria Cybele* in Mississippi." *Journal of Research on the Lepidoptera* 5 (4): 253–54.

————. 1975. "*Amblyscirtes Carolina* and *A. Reversa* (Hesperiidae) in Mississippi and Georgia." *Journal of the Lepidopterists' Society* 29 (3): 177–79.

————. 1995. "Survey of Mississippi Lepidoptera." *Southern Lepidopterists' News* 17 (1): 2–4.

Mather, Bryant, and E. Dingus. 1994. "Field Checklist: Butterflies of Mississippi." Mississippi Museum of Natural History, Jackson.

Mather, Bryant, and K. Mather, 1958. "The Butterflies of Mississippi." *Tulane Studies in Zoology* 6: 62–109.

McKenna, Duane D., and Katherine M. McKenna. 2001. "Mortality of Lepidoptera along Roadways in Central Illinois." *Journal of the Lepidopterists' Society* 55 (2): 63–68.

Metzler, Eric H., John A. Shuey, Leslie A. Ferge, Richard A. Henderson, and Paul Z. Goldstein. 2005. *Contributions to the Understanding of Tallgrass Prairie-Dependent Butterflies and Moths (Lepidoptera) and Their Biogeography in the United States.* Columbus: Ohio Biological Survey, Ohio Department of Natural Resources, Division of Wildlife.

Milne, Lorus, and Margery Milne. 1980. *The Audubon Society Guide to North American Insects and Spiders.* New York: Alfred A Knopf, Inc.

Minno, Marc C., and Jeffrey R. Slotten. 2011. "Inland Colonies of the Eastern Pygmy Blue (*Brephidium Psuedofea Insularus*) along the St. Johns River, Florida." *Southern Lepidopterists' News* 33 (1): 6–8.

Minno, Marc C., and Maria Minno. 1999. *Florida Butterfly Gardening: A Complete Guide to Attracting, Identifying, and Enjoying Butterflies of the Lower South.* Gainesville: University Press of Florida.

———. 2006. "Conservation of the Arogos Skipper, *Atrytone arogos arogos* (Lepidoptera: Hesperiidae) in Florida." 219–22. In *Land of Fire and Water: The Florida Dry Prairie Ecosystem: Proceedings of the Florida Dry Prairie Conference, October 5–7, 2004, Sebring,* ed. Reed F. Noss. DeLeon Springs, FL: E. O. Painter Printing.

Minno, Marc C., Jeffrey R. Slotten, and Mary Ann Friedman. 2006. "Rediscovery of the Wild Indigo Duskywing Skipper (*Erynnis Baptisiae* Forbes) in Florida." *Southern Lepidopterists' News* 28 (2): 49–50.

Nature Conservancy website. www.nature.org.

Neck, Raymond W. 1996. *A Field Guide to Butterflies of Texas.* Houston: Gulf Publishing Co.

North American Butterfly Association. 1993–2013. *4th of July Butterfly Count Reports.* Morristown, NJ: North American Butterfly Association.

———. 2001. *Checklist and English Names of North American Butterflies,* 2nd ed. Morristown, NJ: North American Butterfly Association.

Oberg, Alcestis Cooky, et al. 2004. *The Butterflies of Galveston County.* Dickinson, TX: Texas AgriLife Extension Services.

Ogard, Paulette H., and Sara C. Bright. 2010. *Butterflies of Alabama.* Tuscaloosa: University of Alabama Press.

———. 2011a. "Gorgone Checkerspots in Georgia." *Southern Lepidopterists' News* 33 (2): 41–42, 75–76.

———. 2011b. "Gorgone Checkerspots in Georgia Revisited." *Southern Lepidopterists' News* 33 (4): 156.

———. 2011c. "Searching for Hessel's Hairstreaks in Alabama." *Southern Lepidopterists' News* 35 (2): 61–62, 64.

Opler, Paul A., and George O. Krizek. 1984. *Butterflies East of the Great Plains.* Baltimore: Johns Hopkins University Press.

Opler, Paul A., Kelly Lotts, and Thomas Naberhaus, coordinators, 2010. "Butterflies and Moths of North America." Big Sky Institute, Bozeman, MT. www.butterflies andmoths.org.

Pelham, Jonathan P. 2008. *Catalogue of the Butterflies of the United States and Canada.* Vol. 40 of *Journal of Research on the Lepidoptera.*

Pyle, Robert M. 1974. *Watching Washington Butterflies: An Interpretive Guide to the State's 134 Species....* Seattle: Seattle Audubon Society.

———. 1981. *The Audubon Society Field Guide to North American Butterflies.* New York: Alfred A. Knopf, Inc.

———. 1992. *Handbook for Butterfly Watchers.* Boston: Houghton Mifflin Co.

Raney, Hershel. 2008. "Butterflies of Arkansas." www.hr-rna.com/RNA/Butterfly%20 main.htm.

"Red Slough Wildlife Management Area." USDA Forest Service website. www.fs.usda .gov/detail/ouachita/landmanagement/?cid=STELPRDB5090471.

Ritland David B., and Lincoln P. Brower. 1991. "The Viceroy Butterfly Is Not a Batesian Mimic." *Nature* 350: 497–98.

Ross, Gary Noel. 1994. *Gardening for Butterflies in Louisiana.* Baton Rouge: Louisiana Department of Wildlife and Fisheries, Natural Heritage Program.

———. 1995. "Butterfly Wrangling in Louisiana." *Natural History* 104 (5): 36–43, 86–87.

———. 1996. "Orangetips Die Hard in Louisiana." *American Butterflies* 4 (4): 4–10.

———. 1998. "Monarchs Offshore in the Gulf of Mexico." *Holarctic Lepidoptera* 5 (2): 52.

———. 2001a. "Brainy Butterflies." *News of the Lepidopterists' Society* 43 (2): 43, 47.

———. 2001b. "Monarch Magic." *Louisiana Wildlife Federation* 29 (4): 13–41.

———. 2001c. "New Host Plant for the Silvery Checkerspot, *Chlosyne Nycteis.*" *News of the Lepidopterists' Society* 43 (4): 101, 105.

———. 2003. "Trees Associated with Louisiana Butterflies." *News of the Lepidopterists' Society* 45 (1): 28–30.

———. 2005a. "Life History of the Seminole Crescent." *Holarctic Lepidoptera* 9 (1/2): 1–30.

———. 2005b. "Wise Guys." *Natural History* 115 (8): 72.

———. 2008. "Diana's Mountain Retreat." *Natural History* 117 (2): 24–26, 28, 72.

———. 2009. "Orangetips and Marsh Cattle: An Adventure in Louisiana's Cajunlands." *Southern Lepidopterists' News* 31 (4): 151–60.

———. 2010a. "An Adventure with a Louisiana Swamp Muse." *Southern Lepidopterists' News* 32 (3): 106–19.

———. 2010b. "The Monarch's Trans-Gulf Express: 'A Clockwork Orange.'" *Southern Lepidopterists' News* 32 (1): 11–24.

———. 2010c. "A Tale of Homing in Zebra Heliconians." *Association for Tropical Lepidoptera: Notes,* June, pp. 1–2.

———. 2011. "Diana Still Reigns in Arkansas: A Natural History Essay." *News of the Lepidopterists' Society* 53 (4): 116–23.

———. 2012. "Mt. Magazine and Diana Fritillaries: A Photo Essay." *Southern Lepidopterists' News* 34 (2): 88–104.

———. 2013a. "The Butterflies of Global Wildlife Center in Folsom, Louisiana." *Southern Lepidopterists' News* 35 (3): 116–22.

———. 2013b. "Sightings of Two Subtropical Pierid Butterflies Increasing in South Louisiana." *Southern Lepidopterists' News* 35 (4): 207–10.

———. 2013c. "Wintering of the Gulf Fritillary in South Louisiana." *Southern Lepidopterists' News* 35 (1): 19–21.

———. 2016. "Louisiana Premiers Pollination Celebration." *Butterfly Gardener* 21 (4): 12–15.

———. 2017. "Pollinators Get a Helping Hand from BASF." *Southern Lepidopterists' News* 39 (1): 7–24.

Ross, Gary Noel, and Craig W. Marks. 2002. "Another Interspecific Hybrid between Two *Limenitis* Sp. in Louisiana." *News of the Lepidopterists' Society* 44 (4): 112–14.

Ross, Gary N., and Edward N. Lambremont. 1963. "An Annotated Supplement to the State List of Louisiana Butterflies and Skippers." *Journal of the Lepidopterists' Society* 17 (3): 148–58.

Ross, Gary Noel, and Frances Welden. 2003. "Southern Louisiana." In Gardens and Habitat Program, Regional Garden Guides, North American Butterfly Association, www.naba.org/ftp/solo.pdf.

Salvatto, M. 2011. "Bay Skippers Survey (11 to 17 September 2011): Summary Report." Vero Beach: South Florida Ecological Services Office.

Schweitzer, D. F., M. C. Minno, and D. L. Wagner. 2011. *Rare, Declining, and Poorly Known Butterflies and Moths (Lepidoptera) of Forests and Woodlands in the Eastern United States.* US Forest Service, Forest Health Technology Enterprise Team, FHTET-2011-01.

Scott, James A. 1986. *The Butterflies of North America: A Natural History and Field Guide.* Stanford, CA: Stanford University Press.

Shapiro, Arthur M., and Timothy D. Manolis. 2007. *Field Guide to Butterflies of the San Francisco Bay and Sacramento Valley Regions.* Berkeley: University of California Press.

Shuey, John A. 1989. "The Morpho-Species Concept of *Euphyes Dion* with the Description of a New Species (Hesperiidae)." *Journal of Research on the Lepidoptera* 27 (3–4): 160–72.

Spencer, Lori A. 2014. *Arkansas Butterflies and Moths.* 2nd ed. Fayetteville: University of Arkansas Press.

Stichter, Sharon. 2015. *The Butterflies of Massachusetts.* Middletown, DE: CreateSpace Independent Publishing Platform.

Strickland, Gayle T. 1972. "Recent Studies of Louisiana Rhopalocera." Unpublished manuscript. Presented at annual meeting of Lepidopterists' Society, June 22–25, Trinity University, San Antonio, TX.

Strickland, Gayle, and Jeanell Strickland. 2005. "Butterflies of the Genus Brephidium (Pygmy-Blues)." public.fotki.com/gstrick3/butterflies_of_the/.

Trahan, Jeff. 2009a. "The Butterflies of Caddo Parish, Louisiana." *Southern Lepidopterists' News* 31 (2): 54–59.

———. 2009b. "The Butterflies of Caddo Parish, Louisiana." www.jtrahan.com/butter flies/.

———. 2009c. "First State Record of White Angled Sulphur in Louisiana." *Southern Lepidopterists' News* 31 (3): 89–90.

———. 2013. "First Louisiana State Record of Bordered Patch." *Southern Lepidopterists' News* 35 (2): 80.

Trahan, Jeff, and Terry Davis. 2010. "First State Record of Lyside Sulphur in Louisiana." *Southern Lepidopterists' News* 32 (1): 6.

Tuttle, James P. 1997. "Season Summary of 1996." *News of the Lepidopterists' Society* 39 (2): 1–74.

———. 1999. "Season Summary of 1998." *News of the Lepidopterists' Society* 41 (S1): 1–98.

———. 2000. "Season Summary of 1999." *News of the Lepidopterists' Society* 42 (S1): 1–78.

———. 2001. "Season Summary of 2000." *News of the Lepidopterists' Society* 43 (S1): 1–74.

———. 2002. "Season Summary of 2001." *News of the Lepidopterists' Society* 44 (S1): 1–74.

———. 2003. "Season Summary of 2002." *News of the Lepidopterists' Society* 45 (S1): 1–74.

———. 2004. "Season Summary of 2003." *News of the Lepidopterists' Society* 46 (S1): 1–86.

———. 2005. "Season Summary of 2004." *News of the Lepidopterists' Society* 47 (S1): 1–94.

———. 2006. "Season Summary of 2005." *News of the Lepidopterists' Society* 48 (S1): 1–106.

———. 2007. "Season Summary of 2006." *News of the Lepidopterists' Society* 49 (S1): 1–118.

———. 2008. "Season Summary of 2007." *News of the Lepidopterists' Society* 50 (S1): 1–146.

———. 2009. "Season Summary of 2008." *News of the Lepidopterists' Society* 51 (S1): 1–118.

———. 2010. "Season Summary of 2009." *News of the Lepidopterists' Society* (Apr.): 1–136.

Tveten, John, and Gloria Tveten. 1996. *Butterflies of Houston and Southeast Texas.* Austin: University of Texas Press.

Tyler, Hamilton A. 1975. *The Swallowtail Butterflies of North America.* Healdsburg, CA: Naturegraph Publishers.

US Fish and Wildlife Service website. www.fws.gov.

USDA Forest Service. Forests of the South. www.fs.usda.gov/main/r8/home.

Venable, Rita. 2014. *Butterflies of Tennessee.* Franklin, TN: Maywood Publishing.

Vidrine, Malcolm F., Charles M. Allen, and H. D. Guillory. 1992. "Flight Records of Papilionoidea in Southwestern Louisiana." *Louisiana Environmental Professional* 9 (1): 54–66.

Vidrine, Malcolm F., Charles M. Allen, Bruno Borsari, and Larry Allain. 2001. *Lepidopteran and Odonate Communities in the Cajun Prairie Ecosystem in Southwestern Louisiana.* Proceedings of the 17th North American Prairie Conference. 206–14.

Warren, A. D., K. J. Davis, N. V. Grishin, J. P. Pelham, and E. M. Stangeland. 2012. Interactive Listing of American Butterflies. http://www.butterfliesofamerica.com/list.htm (accessed April 28, 2017).

Wauer, Roland H. 2006. *Finding Butterflies in Texas: A Guide to the Best Sites*. Boulder, CO: Johnson Books.

Williams, Stephen G. 1990. "The Butterflies and Skippers of Southeast Texas: An Annotated Checklist." Unpublished.

NOTES ON PHOTOGRAPHS

In 2013 I began to take pictures of live butterflies in the field, and I chose to use in this volume primarily photographs of live butterflies when available. This decision had nothing to do with any debate concerning the virtues versus the evils of collecting, but rather was made for the simple reason that most people using this book will be looking at individual butterflies outdoors. It made sense to try and match what is in this book to what is encountered in nature.

Because some species have been seen in Louisiana only on a few occasions or not for many years, there were no photographs of in-state living examples available. I therefore reached out to friends in surrounding states to draw from their photographs of these species. The reader may be assured that any photograph in this book taken of a live butterfly outside of Louisiana reflects what that species looks like in the state.

Unfortunately, many butterflies (such as hairstreaks) and skippers (such as grass skippers) rarely present dorsal views, which can be important in identifying a particular species. Therefore, I decided to supplement the numerous photographs of live butterflies with pictures of mounted specimens. I have also used photographs of mounted specimens to reflect some of the variations and subspecies discussed. The photographs of mounted specimens were almost exclusively taken by me, and primarily reflect specimens that are within my private collection.

Some of the photographs reflect specimens from my collection that are maintained in Riker mounts. As a boy, I learned how to spread and mount butterflies using Riker mounts, and I continue to use that method for storage/size considerations and for better protection against carpet beetles. One drawback is that smaller butterflies are sometimes distorted by the pressure of the covering glass, and while I have attempted to avoid including photos of any specimens so affected, in a couple of instances the limited number of specimens available required that I do so.

PHOTO CREDITS AND INFORMATION

I have attempted to identify the photographer of any image I did not take. I have also given locations and dates. Unfortunately, some of my early records were lost in a move, and as a result, in some instances I was only able to give the month and/or year of collection.

From Louisiana, Jeff Trahan, Rosemary Seidler, Vicky Lefever, Jonathan Clark, and Dave Patton contributed their photographs. To the east, from Florida, Mary Ann Friedman and Linda Cooper sent me photographs of butterflies relatively common there but less so within this state. Sara Bright and John Dole also sent me several pictures from the east, including specimens from Alabama, Georgia, Mississippi, North Carolina, and Tennessee. And from the north and west, Herschel Raney, Bob Harden, Tom Lewis, and Cheryl and Norm Lavers contributed photographs of butterflies they have seen in Arkansas, while Bryan Reynolds graciously allowed me to use one of his photos from Oklahoma. I wish to express my sincere thanks to all for allowing me to use their excellent photographs.

In the list below, I have identified the photographer by their initials: Sara Bright (SB), Jonathan Clark (JC), Linda Cooper (LC), John Dole (JD), Mary Ann Freidman (MAF), Bob Harden (BH), Cheryl and Norm Lavers (C/NL), Vicky Lefever (VL), Tom Lewis (TL), Dave Patton (DP), Herschel Raney (HR), Bryan Reynolds (BR), Rosemary Seidler (RS), Jeff Trahan (JT), and my own (CM).

INTRODUCTION

Brazilian Skipper egg, CAD, Sept 14, 2016 (JT)
Red-spotted Purple egg, CAD, July 30, 2014 (JT)
Pipevine Swallowtail egg, CAD, Apr 30, 2012 (JT)
Falcate Orangetip egg, CAD, Apr 14, 2006 (JT)
Giant Swallowtail caterpillar, CAD, Sept 24, 2016 (JT)
Frosted Elfin caterpillar, CAD, Mar 22, 2012 (JT)
Gulf Fritillary caterpillar, CAD, Oct 2, 2009 (JT)
Pipevine Swallowtail caterpillar, CAD, May 7, 2006 (JT)
Monarch chrysalis, CAD, Sept 18, 2016 (JT)
Pipevine Swallowtail chrysalis, CAD, Apr 30, 2012 (JT)
Common Buckeye chrysalis, CAD, Sept 25, 2012 (JT)
Sleepy Orange chrysalis, CAD, July 2, 2007 (JT)
Sleepy Orange freshly eclosed adult, CAD, July 2, 2007 (JT)

IDENTIFICATION KEYS

Butterfly (Eastern Comma) antennae (JT)
Skipper (Wild Indigo Duskywing) antennae (JT)
Moth (Luna Moth) antennae (JT)
Moth (Many-lined Carpet Moth) antennae (JT)
Spread-wing (Southern Cloudywing) profile (JT)
Jet-fighter (Broad-winged Skipper) profile (JT)
Rosy Maple Moth, Dorsal, BOS, July 19, 2010 (JT)
Dorantes Longtail, Dorsal, Polk County, FL, Nov 29, 2004 (LC)
Funereal Duskywing, Dorsal, NAT, June 12, 2009 (JT)
Yehl Skipper, Dorsal, CAD, May 21, 2008 (JT)

Dusted Skipper, Ventral, NAT, Apr 8, 2015 (JT)

Yucca Giant-Skipper, Dorsal, Henderson County, TX, Mar 22, 2016 (JT)

Giant Swallowtail, Dorsal, Clark Creek Nature Area, MS, Aug 7, 2011 (CM)

Zebra Swallowtail, Dorsal, EVA, Mar 25, 2016 (DP)

Checkered White, Dorsal female, DES, May 26, 2014 (JT)

Great Southern White, Male, Polk County, FL, Jan 18, 2012 (LC)

Sleepy Orange, Ventral, Polk County, FL, Oct 30, 2006 (LC)

Cloudless Sulphur, Ventral female, LA, Nov 4, 2014 (DP)

Great Purple Hairstreak, Ventral, CAD, Sept 28, 2014 (DP)

Spring Azure, Ventral, CAD, Apr 16, 2009 (JT)

Harvester, Ventral, CAD, June 8, 2003 (JT)

Little Metalmark, Dorsal, VER, May 19, 2013 (JT)

Gorgone Checkerspot, Dorsal, Stone Road Glade, AR, May 28, 2016 (JT)

Red Admiral, Dorsal, Winchester County, NY, June 24, 2011 (CM)

Eastern Comma, Dorsal, SMN, Aug 20, 2012 (DP)

Viceroy, Dorsal, LAFA, June 5, 2011 (CM)

Tawny Emperor, Ventral male, CAD, Sept 21, 2012 (JT)

Goatweed Leafwing, Dorsal male, SMN, Mar 27, 2014 (DP)

American Snout, Ventral, LA, Nov 20, 2014 (DP)

Monarch, Ventral, LA, Oct 26, 2014 (JT)

Gulf Fritillary, Ventral, NAT, Oct 15, 2014 (JT)

Northern Pearly-eye, Ventral, EVA, Sept 6, 2015 (DP)

Georgia Satyr, Ventral, VER, Aug 24, 2014 (DP)

SPREAD-WING SKIPPERS

Silver-spotted Skipper, Ventral and dorsal, CAD, Sept 12, 2008, and Sept 5, 2008 (JT)

White-striped Longtail, Ventral, NAT, Aug 10, 2012 (JT); dorsal, SMN, Oct 1, 2013 (DP)

Long-tailed Skipper, Ventral and dorsal, BOS, Oct 3, 2012 (JT)

Dorantes Longtail, Ventral, CAD, Oct 5, 2010 (JT); dorsal, Lower Rio Grande Valley, TX, Nov 12, 2007 (C/NL)

Hoary Edge, Ventral and dorsal, CAD, July 9, 2010, and June 27, 2007 (JT)

Northern Cloudywing, Ventral, NAT, Apr 5, 2013 (JT); dorsal, CAD, Apr 11, 2008 (JT)

Southern Cloudywing, Ventral and dorsal, NAT, Apr 14, 2012, and June 19, 2010 (JT)

Confused Cloudywing, Ventral, CAD, Apr 13, 2011 (JT); dorsal, CAD, July 9, 2009 (JT)

Hayhurst's Scalloped-wing, Ventral, CAD, Sept 13, 2007 (JT); dorsal, SMN, July 18, 2009 (JT)

Sleepy Duskywing, Ventral and dorsal, NAT, Apr 5, 2013, and Mar 12, 2012 (JT); ventral and dorsal (male), Apr 6, 2008, Harold Alexander WMA, Sharp County, AR (C/NL)

Juvenal's Duskywing, Ventral and dorsal female, NAT, Apr 5, 2013 (JT); dorsal male, CAD, Feb 28, 2017 (JT)

Horace's Duskywing, Ventral, DES, Aug 5, 2008 (JT); dorsal female and dorsal male, CAD, June 30, 2009, and June 30, 2009 (JT)

Mottled Duskywing, Dorsal, Okaloosa County, FL, June 17, 2005 (MAF); dorsal, AR, Aug 3, 2004 (C/NL)

Zarucco Duskywing, Ventral, CALD, July 27, 2013 (CM); dorsal, CALD, July 27, 2013 (JT); dorsal, CAD, May 15, 2016 (JT)

Funereal Duskywing, Ventral, CAD, Aug 12, 2007 (JT); dorsal, NAT, June 12, 2009 (JT)

Wild Indigo Duskywing, ventral and dorsal, BOS, July 19, 2010, and May 17, 2012 (JT); dorsal and ventral, Washington County, FL, June 6, 2009 (MAF)

Common Checkered-Skipper, Ventral and dorsal male, BOS, July 9, 2006, and June 17, 2012 (JT); dorsal female, GRA, July 19, 2008 (JT)

Tropical Checkered-Skipper, Ventral, CAM, Aug 19, 2012 (JT); dorsal female, SMN, May 5, 2012 (JT); dorsal male, RAP, Aug 16, 2008 (JT)

Common Sootywing, Ventral, Northampton, MA, June 4, 2006 (RJS); dorsal, Clifton Park, NY, June 21, 2006 (JT)

GRASS SKIPPERS

Swarthy Skipper, Ventral, VER, May 19, 2013 (JT); dorsal, GRA, Oct 6, 2012 (JT)

Neamathla Skipper, ventral, Goethe SP, FL, March 21, 2009 (LC); dorsal, Okeechobee, FL, July 13, 2008 (LC)

Clouded Skipper, Ventral and dorsal, CAD, Oct 11, 2012, and Sept 4, 2006 (JT)

Least Skipper, Ventral, SMN, May 5, 2012 (JT); dorsal, CAD, July 29, 2012 (JT)

Southern Skipperling, Ventral, RDR, Oct 8, 2012 (JT); dorsal, CAM, Aug 20, 2011 (JT)

Fiery Skipper, Ventral and dorsal male, CAD, May 27, 2011, and June 26, 2005 (JT); ventral and dorsal female, CAD, Sept 29, 2013, and Oct 13, 2010 (JT)

Cobweb Skipper, Ventral and dorsal male, Bell Slough, AR, Apr 2, 2006 (TL); ventral female, Bell Slough, AR, Apr 26, 2003 (TL)

Meske's Skipper, Ventral, NAT, June 12, 2009 (JT); dorsal male, Walton County, FL, Oct 14, 2006 (MAF); dorsal female, June 19, 2010 (JT)

Sachem, Ventral and dorsal female, CAD, Sept 11, 2011, and Oct 20, 2008 (JT); ventral and dorsal male, CAD, Apr 1, 2012, and June 1, 2012 (JT)

Tawny-edged Skipper, Ventral male, STA, Aug 24, 2014 (JT); dorsal male, Okaloosa County, FL Aug 22, 2016 (MAF); dorsal female, CAD, May 25, 2005 (JT)

Crossline Skipper, Ventral male, CAD, May 2, 2012 (JT); dorsal male, NAT, Apr 26, 2009 (JT); ventral and dorsal female, Stone Road Glade, AR, May 28, 2016 (JT)

Whirlabout, Ventral pair, RAP, Sept 2, 2015 (JT); dorsal female, Santa Rosa, FL, Oct 20, 2013 (MAF); dorsal male, CAD, Sept 5, 2008 (JT)

Southern Broken-Dash, Ventral and dorsal, CAD, May 30, 2012, and Aug 27, 2013 (JT)

Northern Broken-Dash, Ventral and dorsal, CAD, Sept 24, 2008, and Sept 17, 2008 (JT); ventral, LAFA, Oct 5, 2013 (DP)

Little Glassywing, Ventral and dorsal female, BOS, Apr 28, 2011 (JT); dorsal male, SLA, Apr 23, 2011 (JT); ventral female, BOS, April 28, 2011; ventral, CAD, Oct 3, 2016 (JT)

Arogos Skipper, Ventral, dorsal male, dorsal female, and ventral, Okaloosa County, FL, Aug 4, 2009, Aug 27, 2008, Aug 19, 2008, and Aug 19, 2008 (MAF)

Delaware Skipper, Ventral, SLA, Aug 29, 2010 (JT); dorsal male, Arnold Air Force Base, TN, June 13, 2014 (JT); dorsal showing dark scaling, Okaloosa County, FL, Aug 11, 2006 (MAF); dorsal female with dark scaling, SLA, Aug 30, 2009 (CM)

Byssus Skipper, Ventral male, Rick Evans, AR, June 8, 2014 (BH); dorsal male, Rick Evans AR, May 30, 2010 (JT); dorsal female, Rick Evans, AR, June 7, 2014 (BH); ventral, Faulkner County, AR, June 7, 2007 (C/NL)

Zabulon Skipper, Ventral female, dorsal female, ventral male, and dorsal male, CAD, Sept 9, 2010, Sept 9, 2010, Aug 19, 2009, and Aug 19, 2009 (JT)

Broad-winged Skipper, Ventral and dorsal, CAD, July 9, 2010, and Sept 10, 2005 (JT)

Yehl Skipper, Ventral female, SLA, June 2, 2012 (JT); dorsal female, ventral male, and dorsal female, CAD, Sept 9, 2007, May 17, 2007, and Sept 11, 2004 (JT)

Aaron's Skipper, Ventral, Okeechobee County, FL, Sept 27, 2009 (JT); dorsal, CAM, Nov 2006, 2013 (DP)

Palatka Skipper, Ventral, Escambia County, FL, Sept 12, 2007 (MAF); dorsal male, STA, 10, 2008, 2016 (CM); ventral female, STA, Nov 2008, 2016 (CM)

Dion Skipper, Ventral and dorsal male, Escambia, FL, Sept 12, 2007 (MAF); dorsal female and ventral, SLA, Aug 29, 2010 (JT)

Bay Skipper, Ventral, dorsal female and dorsal male, CAM, Oct 5, 2011, and Oct 23, 2014 (JT)

Dukes' Skipper, Ventral, dorsal male and dorsal female, SLA, May 22, 2012, Sept 29, 2010, and May 22, 2010 (JT)

Dun Skipper, Ventral and dorsal male, CAD, July 3, 2009, and Sept 16, 2012 (JT); ventral, WEB, July 27, 2011 (JT); dorsal female, CAD, Aug 17, 2008 (JT)

Dusted Skipper, Ventral, WIN, Apr 1, 2011 (JT); ventral, ventral, and ventral, NAT, Apr 5, 2013, Mar 28, 2012, and Apr 8, 2015 (JT)

Pepper and Salt Roadside-Skipper, Ventral, RAP, Mar 24, 2015 (DP); dorsal, CAD, Mar 20, 2008 (JT); dorsal, CAT, Mar 15, 2012 (CM)

Lace-winged Roadside-Skipper, Ventral, CAT, June 2, 2012 (JT); dorsal, CAD, Sept 17, 2008 (JT)

Common Roadside-Skipper, Ventral, Polk County, AR, Aug 1, 2006 (JT); dorsal, Okaloosa County, FL, Aug 7, 2004 (MAF)

Dusky Roadside-Skipper, Ventral, NAT, May 15, 2012 (JT); dorsal, Okaloosa County, FL, Aug 18, 2007 (MAF)

Celia's Roadside-Skipper, Ventral, Hidalgo County, TX, Oct 22, 2012 (JT); dorsal, Bexar County, TX, June 7, 2014 (DP)

Eufala Skipper, Ventral, CAM, Aug 19, 2011 (JT); dorsal, DES, Sept 20, 2009 (JT)

Twin-spot Skipper, Ventral, SLA, Aug 29, 2010 (JT); dorsal, Okeechobee County, FL, Oct 6, 2012 (LC)

Brazilian Skipper, Ventral and dorsal, CAD, July 30, 2012, and Nov 2, 2009 (JT)

Salt Marsh Skipper, Ventral and dorsal, CAM, Aug 19, 2011, and Oct 5, 2011 (JT)

Obscure Skipper, Ventral and dorsal, CAM, Oct 29, 2016, and Oct 6, 2013 (DP)

Ocola Skipper, Ventral, CAD, Sept 16, 2008 (JT); dorsal, NAT, Aug 5, 2009 (JT)

GIANT SKIPPERS

Yucca Giant-Skipper, Ventral, Henderson County, TX, Mar 22, 2016 (RS); dorsal, Henderson County, TX, Mar 22, 2016 (JT)

Strecker's Giant-Skipper, Ventral, NAT, May 1, 2013 (VL); dorsal, NAT, Apr 26, 2009 (JT)

SWALLOWTAILS

Pipevine Swallowtail, Ventral, CAD, July 15, 2009 (JT); dorsal male, NAT, Apr 3, 2010 (JT); dorsal female, NAT, Apr 3, 2010 (JT)

Polydamas Swallowtail, Ventral and dorsal, Rio Grande Valley, TX, Aug 28, 2011, and June 10, 2012 (LC)

Zebra Swallowtail, Ventral, NAT, May 15, 2012 (JT); dorsal, CAD, Mar 2, 2010 (JT)

Black Swallowtail, Ventral, RAP, Aug 16, 2008 (JT), dorsal male and dorsal female, CAD, May 3, 2008, and Mar 1, 2008 (JT); dorsal male ("*pseudoamericus*" form), Thistlethwaite WMA, May 2, 2010 (CM)

E. Tiger Swallowtail, Ventral and dorsal, CAD, Sept 20, 2011, and Sept 18, 2016 (JT); ventral black female and dorsal female, CAD, May 29, 1993, and Sept 19, 2005 (JT)

Spicebush Swallowtail, Ventral, CAT, June 12, 2012 (JT); dorsal male, CAD, Aug 19, 2012 (JT); dorsal female, NAT, Aug 5, 2009 (JT)

Palamedes Swallowtail, Ventral and dorsal, NAT, Apr 5, 2013, and Mar 12, 2012 (JT)

Giant Swallowtail, Ventral, CAD, June 20, 2007 (JT); dorsal, CALD, July 27, 2013 (JT)

WHITES AND SULPHURS

Common Checkered White, Ventral and dorsal male, CAD, Apr 25, 2012, and Nov 5, 2011 (RS); ventral female, CAD, May 2, 2012 (JT); dorsal female, CAM, Apr 29, 2012 (JT)

Cabbage White, ventral, Steamboat Springs, CO, July 2, 2002 (TL); dorsal female, San Francisco, CA, July 19, 2007 (RS); dorsal male, Shenandoah NP, VA, June 20, 2011 (RS)

Great Southern White, ventral male, CAM, Apr 29, 2006 (JT); dorsal male, Merritt Island NWR, FL, Jan 18, 2012 (LC); ventral female, CAM, Aug 20, 2011 (JT); dorsal female, CAM, July 13, 2013 (CM)

Florida White, Dorsal male, Alta Cima, Mexico, Nov 3, 2007 (RS); ventral, Monroe County, FL, Sept 4, 2013 (SB); ventral female, Miami-Dade County, FL, June 1, 2008 (LC)

Falcate Orange-tip, Ventral male, BOS, Mar 12, 2011 (JT); dorsal female, CAD, Mar 25, 2006 (JT)

Clouded Sulphur, Ventral, Shady Valley, TN, Aug 1, 2013 (SB); ventral, female (alba), Ashville, NC, July 15, 2006 (SB); dorsal male, Stone Mountain, NC, July 17, 1997 (CM)

Orange Sulphur, Ventral and ventral female (alba), CAD, June 16, 2012, and Apr 28, 2011 (JT); dorsal male and female, Sharkey County, MS, June 6, 2009 (SB); dorsal female (alba), Jackson County, AL, Apr 14, 2013 (SB); dorsal male (winter form), Dec 6, 2015 (CM)

Southern Dogface, Ventral, CAD, Aug 13, 2005 (JT); ventral (rosa), Craighead County, AR, Sept 20, 2009 (C/NL); dorsal female and male, Lawrence County, AL, May 30, 2013 (SB)

White Angled-Sulphur, Ventral male, WIN, Sept 30, 2008 (JT)

Cloudless Sulphur, Ventral female, CAD, Aug 29, 2005 (JT); dorsal female, Okaloosa County, FL, Sept 8, 2006 (MAF); ventral male, CAD, July 3, 2009 (JT); dorsal male, Perry County, AL, Sept 8, 2012 (SB)

Large Orange Sulphur, Ventral female, Hidalgo County, TX, Nov 1, 2009 (LC); ventral female (alba), Goliad County, TX, Oct 28, 2016 (JT); ventral male, CAD, Sept 17, 2010 (JT); dorsal male, Monroe County, FL, Sept 14, 2013 (SB)

Orange-barred Sulphur, Ventral male, Okeechobee County, FL, Sept 25, 2009 (JT); dorsal male, Monroe County, FL, Sept 14, 2013 (SB); ventral female, Santa Rosa County, FL, Dec 21, 2008 (MAF)

Lyside Sulphur, Ventral, Cameron County, TX, Oct 25, 2012; ventral, CAD, May 24, 2012 (JT)

Barred Yellow, Ventral (dry/winter form), Baldwin County, AL, Nov 27, 2015 (CM); ventral female (wet/summer form), Osceola County, FL, Sept 26, 2009 (JT); dorsal, Sharkey County, MS, June 4, 2012 (SB); dorsal female, Monroe County, FL, Sept 5, 2013 (SB)

Mexican Yellow, Ventral (summer) Rick Evans, AR, June 9, 2013 (JT); ventral (fall), CAD, Dec 1, 2007 (JT); dorsal male, Patagonia, AZ, Sept 6, 2012 (RS)

Little Yellow, Ventral, BOS, May 17, 2012 (JT); dorsal male, CAD, Sept 29, 2013 (JT); ventral female (alba), Okeechobee County, FL, Oct 6, 2012 (LC); dorsal female (alba), Perry County, AL, Aug 6, 2010 (SB)

Sleepy Orange, Ventral, DES, June 19, 2009 (JT); dorsal, Bell Slough, AR, Sept 29, 2012 (TL); ventral (fall), CAD, Apr 7, 2008 (JT)

Dainty Sulphur, Ventral (summer form), BOS, June 10, 2012 (JT); dorsal, BOS, Aug 11, 2010 (JT); ventral (fall form), NAT, Oct 15, 2014 (JT)

GOSSAMER-WINGED BUTTERFLIES

Harvester, Ventral, CAD, June 8, 2003 (JT); dorsal, Bibb County, AL, June 14, 2012 (SB)

Great Purple Hairstreak, Ventral, DES, Sept 23, 2012, and CAD, July 9, 2007 (JT)

Juniper Hairstreak, Ventral, CAD, Mar 14, 2013, and Mar 20, 2008 (JT)

Frosted Elfin, Ventral, CAD, Feb 27, 2013, and Mar 28, 2010 (JT)

Henry's Elfin, Ventral, Smith County, TX, Mar 10, 2009 (JT); ventral, VER, Mar 16, 2015 (DP)

E. Pine Elfin, Ventral, BOS, Mar 12, 2011, and CAD, Mar 26, 2008 (JT)

Oak Hairstreak, Ventral (No. ssp), CAD, Apr 22, 2012 (JT); ventral (No. ssp), CAD, Apr 22, 2012 (JT); ventral (So. ssp.), SLA, Apr 23, 2011 (JT)

Banded Hairstreak, Ventral, CAD, Apr 25, 2012 (JT); ventral, Hempstead County, AR, May 22, 2011 (JT)

King's Hairstreak, Ventral, Okaloosa County, FL, May 28, 2016 (MAF) and NAT, June 19, 2010 (JT)

Striped Hairstreak, Ventral, Okaloosa County, FL, May 20, 2010 (MAF) and CAD, May 17, 2007 (JT)

Red-banded Hairstreak, Ventral, CAD, Aug 28, 2007, and Mar 25, 2006 (JT)

Dusky-blue Groundstreak, Ventral, Hidalgo County, TX, Nov 3, 2013 (JT); ventral, CAD, Nov 2, 2011 (JT)

Gray Hairstreak, Ventral and dorsal, CAD, Oct 1, 2012, and July 15, 2008 (JT)

Mallow Scrub-hairstreak, Ventral, CAD, Nov 18, 2007 (JT); ventral, Hidalgo County, TX, Nov 4, 2013 (JT)

White M Hairstreak, Ventral, CAD, Aug 12, 2007 (JT); ventral, LAS, Mar 9, 2017 (JC)

Cassius Blue, Ventral, Mission TX, Oct 30, 2010 (JT); dorsal, Okaloosa County, FL, Nov 23, 2012 (MAF)

Marine Blue, Ventral, Hidalgo County, TX, Nov 5, 2013 (JT); ventral, Brewster County, TX, May 20, 2015 (DP)

W. Pygmy Blue, Ventral and dorsal, CAM, Oct 5, 2011, and Aug 20, 2011 (JT)

E. Pygmy Blue, Ventral, Monroe County, FL, July 23, 2004 (JT); dorsal, Monroe County, FL, Jan 24, 2007 (SB)

E. Tailed Blue, Ventral, CAD, Aug 25, 2008 (JT); dorsal female, Okaloosa County, FL, June 7, 2012 (MAF); dorsal male, CAD, Apr 16, 2009 (JT)

Spring Azure, Ventral, CAD, Mar 4, 2009 (RS); dorsal male, Okaloosa County, FL, Apr 20, 2009 (MAF); ventral, Hot Spring County, AR, Mar 18, 2010 (C/NL)

Summer Azure, ventral, CAD, Aug 30, 2009 (JT); dorsal male, Walker County, GA, Apr 19, 2012 (SB); dorsal female, Hamilton County, AL, Aug 15, 2013 (SB)

Ceraunus Blue, Ventral, dorsal female and dorsal male, CAD, Oct 1, 2012, and Nov 11, 2013 (JT)

Reakirt's Blue, Ventral and dorsal female, CAD, May 28, 2009, and Nov 16, 2007 (JT)

METALMARKS

Little Metalmark, Ventral and dorsal, VER, Aug 24, 2014 (DP)

BRUSHFOOTED BUTTERFLIES

American Snout, Ventral and dorsal, CAD, Sept 23, 2012, and Sept 9, 2012 (JT)

Monarch, Ventral, CAD, Aug 30, 2009 (JT); dorsal male, CAD, Nov 21, 2007 (JT)

Queen, Ventral and dorsal, CAD, Nov 25, 2016, and Nov 8, 2012 (JT)

Gulf Fritillary, Ventral and dorsal male, CAD, Sept 5, 2008, and Aug 8, 2005 (JT); dorsal "blonde" male, LAFA June 27, 2009 (CM).

Zebra Longwing, Ventral and dorsal, CLA (escapee from conservatory), Sept 14, 2002 (JT)

Variegated Fritillary, Ventral and dorsal, CAD, June 17, 2008 (JT); ventral (winter form), CAD, Feb 24, 2008 (JT)

Diana, Ventral female, dorsal female and dorsal male, Hempstead County, AR, June 1, 2008 (RS), June 9, 2013, and May 22, 2011 (JT)

Bordered Patch, Ventral, Hidalgo County, TX, Oct 27, 2012 (JT); dorsal, CAD, Oct 21, 2012 (JT)

Gorgone Checkerspot, Ventral, Faulkner County, AR, June 27, 2010 (C/NL); dorsal, VER, May 30, 2013 (JT)

Silvery Checkerspot, Ventral and dorsal, CAD, Apr 16, 2005, and June 1, 2006 (JT)

Phaon Crescent, Ventral and dorsal male, CAD, Aug 6, 2012, and Oct 1, 2012 (JT); dorsal female (top) and male (bottom), DES, Sept 29, 2013 (JT)

Pearl Crescent, Ventral female and male, CAD, Oct 14, 2012 (JT); dorsal male, CAD, Oct 1, 2012 (JT)

Texan Crescent, Ventral "*seminole*," Marianna, FL, June 16, 2012 (MAF); dorsal "*seminole*," LAFA, May 20, 2014 (DP); dorsal "*texana*" male, Hidalgo County, TX, Nov 4, 2013 (JT)

Common Buckeye, Ventral and dorsal, BOS, May 17, 2012, and June 10, 2012 (JT); ventral (rosa), Conway County, AR, Sept 27, 2002 (TL)

White Peacock, Ventral, Okeechobee County, FL, Sept 25, 2009 (JT); dorsal, Hidalgo County, TX, Oct 14, 2006 (JT)

Question Mark, Ventral, BOS, Apr 4, 2012 (JT); ventral (summer), Hempstead County, AR, Sept 23, 2016 (JT); dorsal (fall), CAD, Oct 7, 2010 (JT); dorsal (summer), BOS, Apr 4, 2012 (JT)

Eastern Comma, Ventral, BOS, Aug 10, 2009 (JT); dorsal (summer form), SMN, Aug 20, 2015 (DP); dorsal (fall form), Pulaski County, AR, Oct 24, 2004 (TL)

Mourning Cloak, Ventral, GRA, Feb 15, 2013 (JT); dorsal, CAD, May 22, 2013 (JT)

Milbert's Tortoiseshell, Ventral and dorsal, Lake Owen County, WY, June 23, 1997 (CM)

Red Admiral, Ventral, CAD, May 31, 2006 (JT); dorsal, NAT, Aug 16, 2012 (JT)

Painted Lady, Ventral and dorsal, CAD, Aug 28, 2008, and Aug 28, 2007 (JT)

American Lady, Ventral and dorsal, BOS, May 17, 2012 (JT)

Red-spotted Purple, Ventral, BOS, May 17, 2012 (JT); dorsal, ECA, Aug 6, 2005 (JT); ventral ("*rubidus*" form), SMN, July 29, 2001 (CM)

Viceroy, Ventral and dorsal, CAD, July 7, 2003, and Sept 6, 2013 (JT); dorsal "*watsoni*," LAFA, Sept 26, 2015 (CM)

Common Mestra, Ventral, Hidalgo County, TX, Nov 4, 2013 (JT); dorsal, CAD, Oct 19, 2007 (JT)

Goatweed Leafwing, Ventral male, BOS, Mar 9, 2009 (JT); dorsal female, DES, Sept 21, 2008 (JT)

Hackberry Emperor, Ventral and dorsal males, CAD, Apr 21, 2011, and Apr 30, 2005 (JT); dorsal female (blonde), SMN, 1993 (CM)

Tawny Emperor, Ventral male, CAD, Oct 1, 2012 (JT); ventral female, DES, Sept 24, 2014 (JT); dorsal male, CAD, Oct 1, 2012 (JT); dorsal female and ventral female ("*flora*" form), Alachua County, FL, Aug 20, 2005 (MAF)

Northern Pearly-eye, Ventral, CAT, June 2, 2012 (JT); ventral, May 22, 2011, Hempstead County, AR (JT); *anthedon* antennae, CAT, July 28, 2012 (CM)

Southern Pearly-eye, Ventral and dorsal females, CAD, May 5, 2013, and Oct 16, 2009 (JT)

Creole Pearly-eye, Ventral female, McCurtain County, TX, May 10, 2010 (JT); ventral male, Bibb County, AL, July 17, 2011 (SB)

Appalachian Eyed Brown, Ventral, SMN, Aug 14, 2015 (DP); dorsal, Pine Island, NY, June 30, 2011 (C/NL)

Gemmed Satyr, Ventral, CAD, May 5, 2013 (JT); dorsal, Bell Slough, AR, May 2, 2004 (TL)

Carolina Satyr, Ventral and dorsal, CAD, Apr 9, 2005, and Apr 8, 2009 (JT)

Georgia Satyr, Ventral, RAP, May 30, 2008 (JT); dorsal, Okaloosa County, FL, June 11, 2016 (MAF)

Little Wood-Nymph, Ventral and dorsal, CAD, May 5, 2013, and Apr 18, 2014 (JT)

Common Wood-Nymph, Ventral, NAT, May 15, 2012 (JT); dorsal, Okaloosa County, FL, June 13, 2004 (MAF)

POSSIBLE SPECIES

Olympia Marble, Ventral, Apr 9, 2013, Garland County, AR (C/NL); dorsal, Apr 6, 2013, Lake Catherine SP, AR (C/NL)

Soapberry Hairstreak, Ventral, Chickasaw County, OK (BR)

Edwards' Hairstreak, Ventral, Moore County, NC, June 11, 2001 (SB)

Coral Hairstreak, Ventral, Arnold Air Force Base, TN, June 13, 2014 (JT)

Hessel's Hairstreak, Ventral, Okaloosa County, FL, June 18, 2008 (MAF)

Silvery Blue, Ventral and dorsal, Saline County, AR., Mar 23, 2011 (C/NL)

Baltimore Checkerspot, Ventral and dorsal, Searcy County, AR, May 30, 2007 (C/NL)

Tropical Buckeye, Ventral and dorsal, Hidalgo County, TX, Nov 4, 2009, and Oct 22, 2012 (LC)

Gold-banded Skipper, Ventral, Pulaski County, AR, June 2002 (HR); dorsal, Harold Alexander WMA, AR, May 18, 2007 (C/NL)

Berry's Skipper, Ventral female, Three Lakes WMA, FL, Sept 30, 2009 (LC); ventral male, Okeechobee County, FL, July 10, 2010 (LC); dorsal female, Green Swamp WMA, Sept 28, 2013 (LC)

Palmetto Skipper, Ventral, dorsal male and dorsal female, Okeechobee County, FL, Oct 1, 2006, Sept 28, 2013, and June 22, 2005 (LC)

Carolina Roadside-Skipper, Ventral, Croatan NF, NC, Aug 27, 2006 (JD)

Reversed Roadside-Skipper, Ventral and dorsal, Okaloosa County, FL, Aug 25, 2007, and Mar 25, 2008 (MAF)

Bell's Roadside-Skipper, Ventral, McCurtain County OK, May 10, 2010 (JT); dorsal, Baxter County, AR, May 21, 2008 (C/NL)

QUESTIONABLE SPECIES/RECORDS

Dina Yellow, Ventral, Bauer Hamm, FL, May 31, 2014 (LC)

Julia Heliconian, Dorsal male, Butterfly Festival Conservatory, CLA, Sept 13, 2003 (JT); ventral, Hidalgo County, TX, Oct 21, 2013 (LC)

Great Spangled Fritillary, Ventral, Howard County, AR, June 7, 2014 (JT); dorsal, Lawrence County, AL, Apr 19, 2012 (SB)

Leonard's Skipper, Ventral, Pope County, AR, May 2, 2009 (C/NL); dorsal female and dorsal male, Saline County, AR, Oct 2, 2011 (C/NL)

Common Streaky-Skipper, Dorsal, Starr County, TX, Nov 27, 2009 (LC)

Dotted Skipper, Ventral male, dorsal male and female, Okaloosa County, FL, Apr 9, 2005, Oct 13, 2011, and Aug 20, 2007 (MAF)

INDEX

Note: Only English names of the referenced butterflies and plants are given here. These English names may be correlated with the scientific names in the butterfly checklist starting on page 13 and the appendix of plant names beginning on page 416.